PACKAGING
MANAGEMENT

Packaging functions.

Skins -purposes - protection
                        attractive (mkting)

Materials -cost
Dimensions
Colour.

Packaging

Management

J. H. BRISTON and T. J. NEILL

Gower Press

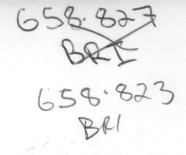

First published in Great Britain by Gower Press Limited
Epping, Essex

ℛ ISBN 0 7161 0146 7

Typeset by the Pentagon Bureau, London
Printed in Great Britain by Biddles Ltd., Guildford, Surrey

iv

FINAL PACKAGING

Contents

v

vi

List of Illustrations

vii

Preface

There has been an enormous growth in packaging during the last two or three decades, with complementary advances in materials and machinery. There has also been a widespread interaction between packaging on the one hand and retail and wholesale distribution on the other. This, together with the increasing integration of packaging with product design or formulation, manufacturing, marketing and distribution means that packaging and its development is of ever growing importance to management.

Up to now, in the UK there have been few books published on packaging, and these have dealt mainly with the technical aspects of the subject. Packaging management, on the other hand, appears to have been completely neglected. This book was written to remedy this deficiency and to cover the important subject of packaging from the managerial point of view.

In addition to dealing with the packaging function and its place in the organisation, modern management approaches to packaging development, specifications, quality control and the use of packaging as a marketing tool are also covered. The chapters on current packaging trends and economics will give the manager an up to date picture of the packaging factors directly affecting him today.

Areas where packaging impinges on society have also been covered. These are assuming even greater importance than some of the marketing or technical aspects, especially to management. The chapters on Solid Waste Management, Packaging and the Law, Education and Training will serve as valuable sources of information to the manager when dealing with problems in these areas.

This book is intended for use by all managers concerned with packaging whether they be located within a packaging supplier, a packaging user, a machinery manufacturer or in one of the many service industries.

1

Introduction to Packaging

There are many definitions of packaging, two widely quoted ones are:

1 'Packaging is the art, science and technology of preparing goods for transport and sale'.
2 'Packaging may be defined as the means of ensuring the safe delivery of a product to the ultimate consumer in sound condition, at the minimum overall cost'. — *& also appealing to consumers*

There is another definition which sets out to explain what packaging is by saying what it does. It is: 'Packaging must protect what it sells, and sell what it protects'. This adds to the first two definitions the important subject of sales appeal.

Packaging, as a subject for study, is of fairly recent origin but in fact the art of packaging is as old as man himself. Possibly the first use of packaging was when primitive man used leaves to wrap uneaten portions of meat. If the tribe was on the move, and uncertain when fresh game would next be encountered, the fact that meat could be carried with them would be important.

For liquids, the use of animal skins as water bags was an equally important packaging innovation. Another early package was the wicker basket while such materials as cloth, paper and wood also made early contributions to packaging. Glass, too, has a long history, one story attributing its accidental discovery to Phoenician traders when they built wood fires for cooking pots, on the sandy beach, using blocks of soda on which to stand their cooking.

Metal was comparatively late on the scene but was, of course, responsible for the enormous market which now exists for processed foods. Plastics were the latest arrivals on the packaging scene and they are still carving out their own particular niche.

Importance of the Packaging Function

The importance of the packaging function should be obvious from the definitions of packaging given above; packaging protects the product and delivers it to the point of sale in sound condition. In addition, it adds sales appeal to the product and so helps to build up sales. If packaging is to perform its proper function, however, it must be considered at as early a stage as possible. This means that packaging must be considered at the design or formulation stage. A simple example is the case of mechanical or electrical equipment.

If the packaging function is fully integrated with that of product design, it will be possible to avoid costly mistakes leading to the production of equipment which needs extremely complex and costly packaging in order to confer adequate protection during its journey from factory to consumer. Sometimes it may be something as simple as a badly positioned lever which necessitates special packaging, but whatever the reason, it will usually be found that more thought at the design stage would have avoided the problem.

At the other end of the chain, packaging is part of marketing and must be considered at the start of any marketing plans. It will be seen that the packaging function is involved with many other functions within the company. The following chapter ('Organisation of the Packaging Function'), outlines some of the alternative operating systems for the packaging function but it will be useful at this stage to consider its position *vis-à-vis* other parts of the company. Efficient communications are important because of the diverse disciplines represented in the packaging operation. They include, *inter alia*, chemistry, physics, engineering, marketing, design, law and accounting. Figure 1.1 shows the departments with which the packaging function is regularly in contact.

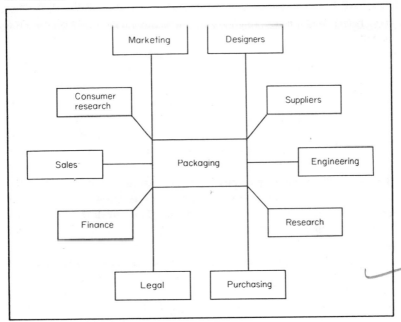

Figure 1.1 Packaging communications

Two of the major functions are purchasing and marketing. The purchasing function is responsible for all packaging purchases and because of this it needs to be familiar with all work on development projects. Liaison with outside suppliers is also the main responsibility of the purchasing function. When personnel are involved in packaging work directly with outside suppliers they need to keep their company buyers fully advised.

Marketing liaison involves keeping those with overall product responsibility in the picture where packaging is concerned. During the initial stages of new product development the marketing and consumer research functions, together with outside designers, work with the packaging team to develop and assess package shape and surface design. Legal experts are called in to advise on the validity of pack copy and on matters affecting trade marks.

3

On the technical side, development of the product and of the packaging and filling equipment involves liaison with the research and engineering functions.

It will be seen, therefore, that to operate efficiently the packaging man needs a knowledge of the disciplines and techniques used in the departments with which he is in regular contact. Equally important he needs to be able to communicate clearly, both verbally and in writing.

The following chapters in this book reflect the complexity of packaging but have been planned to break the subject down into self-contained elements. Thus, Chapter Two, as mentioned earlier, deals with the organisation of the packaging function, with examples being taken from both small and large companies. Two chapters then deal with the approach to packaging development while one on package evaluation outlines the important tests which should be carried out to determine the suitability of the package both in use and at the filling plant. Next, comes a chapter on packaging specifications, followed by a chapter on the measurement of packaging quality dealing with subjects such as methods of measurement, and control systems in user and supplier companies.

After a chapter dealing with packaging economics and cost reduction techniques the use of packaging as a marketing tool is considered. The relationship of packaging to functions such as advertising is discussed together with a brief treatment of package decoration and identification techniques. This chapter also includes the subject of packaging and sales promotion.

The last three chapters deal with special topics of ever-increasing importance; namely, 'Packaging and The Law', 'Solid Waste Management and Packaging', and 'Education and Training'.

Packaging Criteria

The final form of any package is influenced by many factors but logical packaging development can be achieved by considering various packaging criteria.

These are basically five in number; namely,

1 Appearance
2 Protection

3 Function
4 Cost
5 Disposability

These are the main considerations when developing a pack. The relative emphasis placed on them depends on the product and on its marketing requirements. Some examples are listed below:

1 A cosmetic product; the most important criterion likely to be APPEARANCE.
2 An aerosol hair spray; the most important criterion is FUNCTION.
3 Ethical Pharmaceutical; the most important criterion is PROTEC-TION.
4 For a one-trip bottle of milk; the most important criteria are COST and DISPOSABILITY.

This is not to say that the other criteria are ignored. For example, the most exotic cosmetic product must have some cost limitations attached to its packaging requirements and the most humble household product will need to have some attention paid to its appearance (especially if it is sold through self-service stores). One thing is sure and that is that DISPOSABILITY is growing in importance. For this reason a special chapter is devoted to the subject of solid waste management. The remaining four packaging criteria will now be discussed in more detail.

Appearance
ASPECT

This criterion is growing in importance with the growth of supermarkets and cash and carry. The appearance of a package is important for a number of reasons.

1 It has to identify the product throughout the distribution chain and especially when it reaches the final consumer.
2 It may have to carry instructions for use.
3 It may have to carry information about the contents in order to satisfy legal requirements (as with poisons, or with most foodstuffs).
4 It will usually carry the brand name or the name of the manufacturer, or both.

5

5 It can act as an important sales aid.

The appearance of a package is dependent on two main elements; shape and surface decoration. There are often conflicting requirements for package shape. On the one hand, the market requirements may be for a complex shape to fit the product image, whereas the retailer's requirements are for stability in stacking, and efficient use of shelf space. One example of a special limitation on package shape is that bottles containing certain poisons must be ribbed to provide positive identification in the dark (see Chapter Eleven 'Packaging and the Law').

Surface decoration may be achieved either by labelling or by direct printing on the package. The subject of surface decoration is dealt with more fully in Chapter Ten on 'Packaging as a Marketing Tool'.

An important aspect of the package appearance is that it must be durable. In the case of retail goods this means that the appearance must still be attractive enough to sell goods as well as to identify them. In the case of industrial goods, the identification factor is often more important than sales appeal, but even here any deterioration in appearance should be avoided. The function of identification becomes of paramount importance when the subject of packaging for the defence services is considered. If military equipment (ammunition, etc), is wanted in a hurry the packages must be capable of giving instant and positive identification. The other factor is that of time. Packaging for the services is designed for long-life (of the order of five years or more), and the identification must be equally long lasting. The requirements for service packaging appearance can, therefore, be broadly stated as follows:

1 Identification must be clear and positive (no ambiguity).
2 The identification must be long lasting under all possible climatic conditions.

Protection

Although protection may not always be the most important criterion for every packaging situation it is a requirement that is rarely completely absent. The protection required by the product will vary enormously with the nature of the product itself, the final destination, the distri-

6

bution system and the total time that protection is required. Protection is normally required against two main hazards; chemical and physical.

Chemical

Chemical hazards are here taken to include microbiological hazards.

Product/Packaging Material Compatibility Chemical interaction between the product and its container is doubly undesirable. Not only may it lead to undesirable changes in the product but it may cause a weakening of the container with consequent failure in service.

One of the most common examples of product/package incompatibility is packing of acidic or alkaline products in plain tin plate containers. Both types of interaction can be found in this situation. For example, certain coloured fruits are decolourised by the tin coating on tin plate and certain detergents corrode tin plate and eventually cause leakage of the product. The answer is to prevent contact of the product and the tin plate by lacquering of the surface. This solution is more likely to be successful in the case of preventing deterioration of the product since even if pin holes or other imperfections occur in the lacquer, the area of contact will be very small, in relation to the volume of the product.

The presence of pin holes when a corrosive product is present, however, leads to a concentration of attack at a small area of exposed metal and so can easily lead to pinholing of the container followed by product leakage. Great care must be taken, therefore, in lacquering in such instances to ensure that the total area is covered. It will often be necessary to flush-lacquer the completed can even though the can is made from lacquered tin plate.

Some effects of product/package incompatibility are more subtle than the ones just considered. It has been known for glass, which is normally considered to be extremely inert, to affect certain products. Aqueous liquids very slowly acquire an alkaline reaction when stored in glass and this can adversely affect certain alkali-sensitive drugs or transfusion liquids. Specially treated glass (sulphated glass) is available for the packaging of such liquids.

Ingress of Liquids and Vapours The packaging material is frequently called upon to act as a barrier against entry into the package. One of the commonest causes of product deterioration is water, either in the form of liquid water or moisture vapour. Many granular or powdered chemicals, for instance, cake badly under humid conditions, while others become extremely corrosive when damp. Paper is not normally a barrier to moisture vapour but glass, tin plate and aluminium are. Plastics are not 100 per cent barriers but many are sufficiently so for normal uses. It should be remembered, incidentally, that even the best container is only as good as its closure or seal.

The hazard of liquid water ingress may arise when a product is shipped as deck cargo or through climatic conditions, particularly in tropical countries. In such situations it is essential to provide an outer, water-proof package using a plastics film or a varnished fibreboard case. It should be realised, however, that under conditions of fluctuating temperatures it is possible for condensation of moisture vapour to occur within the container, even though it is a barrier to liquid water. When this type of action can occur it is necessary to place a desiccant within the container to absorb any moisture vapour present.

Many products require protection against the ingress of gases, particularly of oxygen. Fatty foods, for instance, become rancid when the fat is oxidised. Also many pharmaceuticals are adversely affected by reaction with oxygen. Once again, the best barriers are glass and metals (subject to the provision of efficient closures). Some plastics have quite low permeabilities to oxygen while others have high ones. The success of materials with an appreciable permeability will depend on external conditions, such as temperature and humidity and on the shelf life required.

Finally, many foodstuffs can be adversely affected by pick-up of external odours or flavours during transit and storage. The package must, therefore, act as a barrier in such cases. Metal and glass are complete barriers but paper, board and plastics are not. Plastics vary a great deal in their permeability to odours and flavours but one of the highest permeabilities to essential oils (which constitute a high proportion of odours or flavours) is possessed by low density polyethylene. This material should not, therefore, be used alone where there is thought to be a real danger of odour or flavour pick-up, particularly if the product itself has only a slight odour or flavour of its own.

Loss of Liquid or Vapour   Loss of liquid or vapour can also lead to adverse changes in the product. Examples include the drying-out of tobacco or cigarettes, loss of solvents from shoe polish (leading to hardening of the product), loss of flavour from foodstuffs and the loss of perfume from cosmetics. The problems are much the same as those discussed above and the types of barrier materials are the same. The provision of an efficient closure is equally important.

Micro-organisms Where the product is a food or pharmaceutical which has to be sterilised (usually by heat) either prior to packaging or in-package, then the function of the package is to prevent the ingress of fresh micro-organisms. Suitable packaging materials are tin plate, aluminium, glass and some plastics.

For dry foods which have not been sterilised, the package has to prevent the ingress of moisture which would then promote microbial growth of organisms already present. The importance of efficient packaging in such instances cannot be over-estimated. Apart from spoilage of the contents, some bacteria produce toxins which can be deadly. It is not only edible products, however, which have to be protected from attack by micro-organisms. Glass is attacked by enzymes produced by certain moulds so that the glass jar and the glass covers for instrument dials are both susceptible to damage. Similarly, aluminium is attacked by an acid produced by mould growth. In addition, outer cases of wood or fibreboard may have to be treated with a fungicide for export to tropical regions because of possible attack by fungi and bacteria.

Physical

Physical hazards in distribution may be static or dynamic and may be summarised under the following headings:

Compression This arises from stacking in transit or in storage. If the primary pack is sufficiently stout (for example, a cylindrical tin plate container with flat ends), the outer pack need do no more than contain the primary packs, which can take the stacking load themselves. For weaker primary packs such as cartons or flexible plastics packs, the outer container must be constructed to take a large proportion of the maximum stacking load likely to be encountered. It must be emphasised that damage to the bottom containers in the stack is not the only danger. More serious is the risk of stack instability, with possible damage to many more containers as well as risk to life or limb.

Impact   Impact damage can arise through dropping of the package or shunting of rail cars, etc. In addition to breakage of containers, giving leakage, there is also the risk of damage to equipment by distortion.

Puncturing   Puncturing can occur through similar hazards to those outlined above for impact. The main risk is leakage of liquid or powdered contents but punctures may also provide inlets for moisture vapour, with consequent corrosion of metallic products.

Vibration   Vibration can cause a multitude of ills from abrasion and scratching of the outside of the containers (perhaps with loss of identification) to breakage of the contents. The package has a vital role to play as a cushion when transporting fragile goods.

Effect of Temperature   The effects of high temperature on a product are usually more serious than those of low temperature. Corrosion effects, for example, are accelerated by high temperature, as are other chemical changes and biological spoilage. It is also necessary to consider the effect of changes in temperature even when the extremes likely to be encountered are not thought to be harmful. Thus, cooling of a warm, moist atmosphere will lead to deposition of some moisture and condensation and this liquid water can then cause corrosion of metal parts or deterioration of water-sensitive chemicals, foodstuffs, etc. Examples of products likely to be adversely affected by an increase in temperature include chocolates, (which soften and melt at elevated temperatures and become unsaleable); fish, (which rapidly develops a strong off-odour and then becomes inedible); frozen foods (which thaw out and then start to deteriorate biologically); and many pharmaceutical products which lose their therapeutic activity (~~or may even become biologically inimical~~) if stored at high temperatures.

In the case of fish, the packaging usually includes a quantity of crushed ice to preserve the contents at low temperature. Boxes are now available made from a foamed plastic — expanded polystyrene — which acts as an efficient heat insulant and allows economies in the use of ice. With deep frozen foods there must be provision for maintaining their low temperature right the way through the distribution chain and at the retail store.

The problem with chocolates is a little different since it is not feasible to keep them under refrigerated conditions during transit and in overseas retail stores. The solution here is to reformulate the chocolates to give a higher melting point when exporting to destinations known to experience long periods at temperatures of around 35° to 40°C.

Falls in temperature are not normally so important and will, in fact, often increase the shelf-life. One important exception is emulsion paints which consist of a dispersion of pigments and synthetic resins in water. Too low a temperature will freeze the water component and thus break the emulsion. Tins of emulsion paint exported to low temperature countries should, therefore, be placed in an outer packaging giving adequate heat insulation. Similar remarks apply to many adhesives as these tend to undergo physical changes at low temperatures; such changes are difficult to reverse by the consumer.

It should be noted that very low temperatures are not only encountered by products exported to cold countries but also by products in the holds of aircraft.

Effect of Light   Light may adversely affect many products. The effect varies from changes to colour, embrittlement of some plastics and catalysis of chemical reactions (such as oxidation of fats, giving rancidity). Many pharmaceutical products are affected by visible or ultra-violet light and must, therefore, be packed in opaque containers or coloured glass bottles.

Macro-organisms   This heading covers both insects and rodents. In addition to the loss of the product actually eaten by these organisms, there is the added loss through contamination of the remaining products. Paper and board do not constitute much of a barrier to the more voracious pests but plastics are quite efficient. Tin plate and glass are even better barriers but a determined rat can penetrate virtually any barrier (including glass and metal).

Pilferage   Although no package is a complete defence against pilferage some can often make the job of the thief much more difficult. One way of reducing pilferage is by 'containerisation' whereby a large number of normal shipping containers are put into a large van-type container. These large containers are solidly built and can be padlocked. The shrink wrapping of whole pallet loads can also be of help in this area.

Function

The functions which a container may be called upon to perform can be divided into two main classes:

1 Those concerned with its end use, and;
2 Those concerned with its behaviour on the packaging/filling line.

End-use performance

This is obviously important since faulty performance will lead to dissatisfaction with the product itself and, hence, to a reduction in sales. End-use package functions include:

1 Display
2 Ease of opening
3 Convenience
4 Dispensing

Display The package may be used as a display item in its own right by means of attractive surface decoration or it may act as a display item for the product. The former has already been dealt with under the heading 'Appearance'. The second factor ie product visibility is not always desirable, of course, especially where the product is sensitive to light. When visibility is required, it is usually an aid to identification or to add sales appeal to the finished pack. The latter is an increasing trend in large chain stores.

In the case of flexible packaging there are many plastics films, together with regenerated cellulose film, which can be obtained in a fully transparent form. For rigid packs the choice lies between glass or certain plastic packs such as PVC bottles. One other possibility exists where the product does not have to be protected from the environment and that is the use of a cutaway or some other form of open container. One example is the use of cotton or plastic nets for the packaging of nuts, oranges, onions, etc. Another, different, example, is the packaging of eggs where a cutaway moulded pulp container can be used which will give sufficient visibility while still maintaining mechanical protection. Until the advent of translucent polystyrene egg packs this was the only way of obtaining product visibility plus mechanical protection.

Ease of Opening   This is a very difficult function to satisfy since it is nearly always combined with the necessity for the pack to maintain its seal or closure integrity until the moment when the customer wants to open it. Tear tapes may be the answer in the case of film overwraps while there are now a number of easy opening devices for metal cans, such as beer cans. The most difficult field is probably that of plastic pouches, especially those containing liquid products. Tearing plastic films is not easy, even with solid products inside but to tear open a plastic pouch of liquid without spilling the contents is extremely difficult. The usual way round this problem is to use a pair of scissors but this is not always a possible solution.

Convenience   The need for convenience in packaging has led to the growth of packs where the package and the product are completely integrated and where it is difficult to separate product and package performance.

An excellent example of this type of product/package integration is the aerosol or pressurised pack. The product as used by the consumer (for example, an insecticidal mist or a shaving foam) is not actually contained by the package but is produced at the moment it is required. The production of the mist, foam, etc, is a function of the complete pack. With the shaving foam, the container has to be strong enough to withstand the high internal pressures generated by the mixture of soap solution and liquefied gas propellant while the valve, part of which is an actuating button, also provides an expansion chamber in which the foam is formed as the pressure is released on the mixture of soap solution and propellant. Perhaps an even better example is the aerosol hair lacquer. At least shaving soaps or creams are available in other forms which do a similar job to the aerosol shave foam, but the hair lacquer market was actually built up from nothing, by the aerosol pack, into a multi-million package business.

Aerosols are only one example of convenience packaging. A number of examples can be quoted from the food industry including boil-in-the-bag food pouches and TV dinners. The boil-in-the-bag pack (the pouch and its contents), are placed into boiling water for a few minutes, the package is removed and the cooked food is dispensed, ready for eating, by cutting open the pouch. This type of pack has to be able to contain the food during transit and storage, resist the temperature of boiling water, and be readily opened when required. The main convenience factors are, absence of cooking smells, absence of the need to clean saucepans after cooking, and ease of preparation of a complete meal (at short notice, if necessary). The TV dinner is somewhat similar in concept. A pre-cooked meal is packaged in an aluminium or plastics tray, then deep frozen. A brief heating period in an oven (hot air for the aluminium tray, micro-wave for the plastics tray) again provides a meal with no preparation, waste or washing-up problems. In both cases the package is an integral part of the product.

Dispensing   This is allied in some ways to ease of opening and convenience. A great number of the advances made in dispensing devices, such as pourers, spouts and taps, have been made with the aid of plastics because of the flexibility of design offered by these materials. Dispensing aids range from a plastic cap which is easily punctured to give a hole through which a liquid can be shaken, to complicated retractable taps for drums which can then be rolled along the ground without breaking-off the tap.

Machine Performance

This can be a very important aspect of container design. Examples of container design which can affect filling speeds or types of filling equipment are neck diameter, stability (relationship between base diameter and diameter at top of container), rigidity of container wall (a certain amount of rigidity is necessary for vacuum filling), and variations in container weight (which affects reproduceability of weight filling).

A change from one type of container to another may necessitate changes in filling equipment ranging from a simple modification, to a complete new filling line. Thus, a luboil line designed to run using full aperture metal cans with double seamed lids, would need major alterations to run on lightweight plastic bottles with narrow necks and screw-on caps. Plastic containers can be (and are) used, however, in the form of full aperture plastics bodies with metal lids fixed to the plastic bodies with slightly modified double seamers. The light weight necessitated modification to the conveyor equipment but the filling could then be carried on as before (at high speeds) because the aperture of the containers was unchanged.

It is difficult to divorce the package from the machinery which is used to fill it and handle it (both prior to and after filling). Package design and machinery design are often inter-dependent and thus the packaging/machinery interface must be taken into account at as early a stage as possible.

In particular, new packaging materials usually pose problems in use on high speed equipment developed primarily for more established materials. The case of low density polyethylene film was an excellent example in this connection. Wrapping equipment had previously been designed to deal with paper or with regenerated cellulose film, both of which are stiff, even in thin gauges. Polyethylene film on the other hand is very limp. The mechanisms designed to move paper and cellulose films through the wrapping machine usually worked on the principle of mechanical fingers, pushing the material along, and new equipment had to be developed before polyethylene film could capture an appreciable share of the wrapping market.

The packaging/machinery interface is sometimes difficult to distinguish. Many modern packaging machines produce filled packages from film or sheet without any intermediate package being formed. These form/fill/seal machines bring package making into the product manufacturer's plant and this can introduce problems. One which can arise is the entry of different unions, such as printing unions, into the factory. Different types of expertise also have to be employed and this may create difficulties for industries such as dairies which are confronted with the problems of plastics technology in the form of bottle blowing. This would be so in the case of the form/fill/seal technique where bottles are blown from plastics granules, filled and heat sealed, all on the one machine.

Cost

One of the definitions of packaging quoted earlier was based on the idea of delivering the product in sound condition at minimum overall cost. Delivery at minimum overall cost is a very important function of packaging but it is essential to be clear what constitutes 'overall' cost. The subject of packaging economics will be dealt with in more detail in Chapter Eight but meanwhile it should be noted that some of the factors contributing to the overall cost of packaging a particular product are:

1 Package cost (delivered to factory).
2 Storage and handling costs of the empty package.
3 Filling cost (including quality control and handling of filled packages).
4 Storage costs of the filled package.
5 Transport costs of delivering filled packages.
6 Insurance costs involved in transport.
7 Losses due to breakage or other spoilage of the product (including loss of goodwill).
8 Effect of the package on sales.

In many cases the purchaser of packaging takes the view that he has achieved an overall saving if he reduces his package costs but the above list should be a salutary reminder that this is a very short-sighted attitude. A simple example would be the case of a plastic container which while more expensive, *per se*, gave savings in freight cost and in a reduced breakage rate.

Sources of Information

Since this book cannot hope to cover the broad technical field of packaging in great detail, some suggested further reading is given here. A valuable source of up-to-date information on packaging developments is the technical packaging press. Apart from the journal issued by the Institute of Packaging ('Packaging Technology') there are journals such as 'Packaging News', 'Packaging', and 'Packaging Review', all of which cover the general packaging field. In addition, there are more specialised journals such as 'The Converter', 'Tin Printer and Box Maker' and 'Converting Industry'. The above are all UK journals but for wider European coverage there are 'Emballages' (France), 'Imballagio' (Italy) and 'Verpackung' (Holland). Among journals covering the North American scene there are 'Modern Packaging' and 'Package Engineering' (USA) and 'Canadian Packaging' (Canada).

Packaging Exhibitions are also useful sources of information and these are held in various countries, including the UK, Germany, France, Holland, Italy and the Scandinavian countries. In addition to obtaining literature from the stands, and talking to company personnel, it is often possible to see new equipment in action.

Finally, the branches of the Institute of Packaging in many parts of the country hold regular meetings at which experts in their fields are only too happy to share their expertise and experience.

2

Organisation of
the Packaging Function

The importance of packaging in many product areas in today's highly competitive market place has been stressed in Chapter One. Packaging is no longer a poor relation in the overall company organisation. It makes a vital contribution to overall profitability in many industries. Surveys carried out in the marketing of retail consumer products have shown that packaging is often the major single reason ascribed for the success or failure of a new product. Similarly, in industrial markets new packaging methods have been adopted to modern distribution channels to increase profitability.

Packaging management is a complex subject. Implementation of the packaging responsibility involves close contact with many different departments. These cover a range of disciplines from scientific and technical to commercial and legal. The degree of involvement with each department varies depending on the type of operation in which the company is involved. It also varies within a given company at any given time depending on the nature of the project in hand. For example, a cost reduction exercise may concern only a single department, say manufacturing, while a new product development may involve twelve different areas of responsibility.

In the last ten years many companies have kept pace with the increasing importance of packaging by establishing separate departments to deal with this function. Such departments are staffed by packaging professionals, fully trained in all aspects of packaging technology and modern management techniques. Obviously this is only justified in companies where packaging has a major effect on the total profitability. However, separate profit accountable packaging departments are still the exception rather than the rule, even in large companies, although they are much more common in the United States of America than in Europe. The existence of separate packaging departments could undoubtedly be justified in many more companies but it is not the answer for every situation. Because of the diverse nature of the packaging activity the most suitable organisation can only be chosen after a study of the overall operation. Many large companies employ more than one approach to packaging organisation in different areas of interest.

The relative position of the packaging function in many companies owes more to history than to logic. In days when packaging was relatively unimportant it was allocated as a minor responsibility to one of the main company departments, for example, purchasing, research or manufacturing. The allocation was made either on the basis of existing expertise within the department, or on the relative degree of involvement of the department in the total packaging operation. As the importance of packaging increased, the particular department, rightly or wrongly, maintained the responsibility. Unfortunately, in many instances, because of such historical reasons, packaging is still regarded as a minor part of a larger department and often is not situated in the best organisational situations.

In this chapter the main methods of approaching the organisation of the packaging function will be studied and examples will be given of the types of companies using these methods. Job descriptions of individual positions within the organisations will also be outlined. The methods will be considered in terms of the packaging function title under the following headings:

1 Packaging department
2 Corporate packaging organisation
3 Packaging committee/co-ordinator
4 Packaging supplier

These will now be considered in detail.

Packaging department

Over the last ten years the main trend in development of packaging organisations has been the formation of separate packaging departments. This trend is due to an increasing management awareness of the ever-growing importance of packaging in many product areas and is particularly relevant to industries with the following features:

1 Relative cost of packaging to product is high; cosmetics, toiletries and speciality goods.
2 Annual expenditure on packaging materials is high; household cleaning products, tobacco and cigarettes and food products.
3 Unit product cost is high; electrical appliances, watches, speciality chemicals and wines/spirits.
4 Large number of packaging items are handled; automobile spare parts, retail store groups and private label products.

The packaging department size will vary with the individual company being considered. The example considered here is that of a large department in a company with a major interest in packaging, for example, a cosmetics company. Figure 2.1 shows an organisation chart for the packaging department in this type of company. The four major areas connected with packaging are all integral parts of the department. This type of organisation is something of an ideal situation and very few companies will have all four areas responsible to the packaging department. It does however describe all the groups which exist in various packaging departments. The responsibilities for the various sections of the department can be considered as follows:

Packaging manager

This is the usual title for the department head. In some instances, where the packaging function reports to the managing director, the title will be 'Packaging Director'. In the USA, 'Vice Presidents of

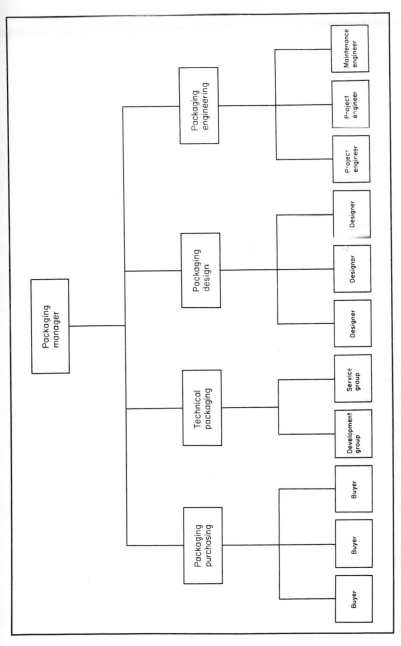

Figure 2.1 Packaging department organisation chart

Packaging', reporting to the chairman or general manager are quite common. In the UK and Europe the packaging managers usually report to the director in charge of purchasing, research, manufacturing or marketing.

The packaging manager is usually an experienced professional with practical experience of at least two of the groups working for him. Often he will also have had experience working in closely associated departments, for example, research, manufacturing or purchasing. In some companies the latter requirement is a pre-requisite for the position. The main responsibilities of the packaging manager are:

1 Overall responsibility for all aspects of packaging. Liaison with other departments and outside suppliers. Drawing up and implementation of departmental policy.
2 Total planning of the work of the department. Co-ordination of the individual units. Allocation of project priorities. Overall control of standards and procedures. Organisation of facilities.
3 Preparation and implementation of annual budgets. Approval of major items of capital expenditure.
4 Personnel administration. Recruitment and training. Staff reporting and development. Evaluation of educational needs and facilities. Responsibility for remuneration policy, organisational structure and job descriptions.
5 Responsible for overall departmental safety and for security of confidential operations.

In a large department such as the one illustrated in Figure 2.1, administrative work will occupy a large proportion of the manager's time. In this situation effective use of modern management methods are vital to ensure an efficient departmental operation. The 'Management By Objectives' approach, outlined in Chapter Three, is an effective method of operation. Under this type of system the individual sections are responsible for the effective functioning of their units. The responsibilities of the four sections as illustrated in Figure 2.1, Packaging Purchasing, Technical Packaging, Packaging Design and Packaging Engineering will now be outlined.

Packaging purchasing

Organisation of the purchasing section is relatively simple. This applies

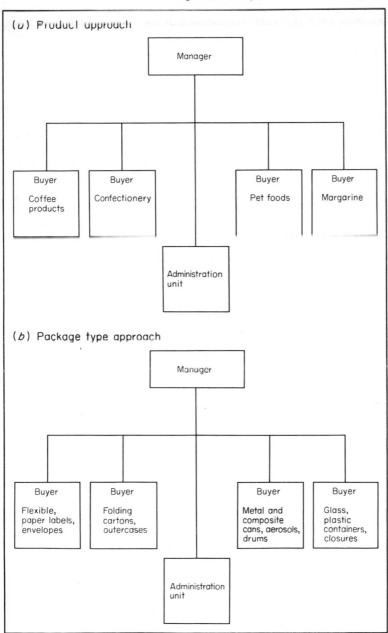

Figure 2.2 Packaging purchasing organisation

whether, as in this case, the unit is part of the packaging department or it is part of a main purchasing department. The section consists of a manager and a team of buyers or purchasing agents. These are supported by an administrative section to progress the purchase orders and to handle day-to-day details.

The number of people in this section will depend on either the total purchasing bill or on the complexity of the items purchased. The buyer's responsibility can be allocated in one of two ways; either by type of product or by type of package. Figure 2.2 illustrates the two possible approaches to organisation in a multi-product food company.

In the first approach the individual buyer purchases a range of packaging items for a given product group. Using this method, the buyer obtains a good grasp of the product type and a general experience of many packaging material types. One of the disadvantages is that the packaging supplier's representatives have to service several buyers within the same company. There is also often duplication of effort within the buying team.

The second, more popular method, has the advantage that the buyer becomes experienced in one particular range of packaging material items. This method is usually adopted by companies where a large number of different products are manufactured, for example, large retail store chains. Companies using this approach often rotate the buyers after a two year spell in each position to broaden their experience.

The main responsibilities of the purchasing section are as follows:

1 Supply of packaging items, to a given specification, to the manufacturing point, by a stated delivery date.
2 Primary contact with packaging material suppliers. Selection and development of suppliers. Financial checks on suppliers' stability Investigations of complaints concerning delivery or service.
3 Determination of exact packaging material costs. Supply of cost information to other departments. Preparation of reports for management. Estimation of future price trends.
4 Inventory control of packaging items. Identification and write-off of obsolete materials.
5 Implementation and execution of cost reduction programmes. Value analysis approach to material cost and supply methods.
6 Preparation of annual budgets for packaging materials.

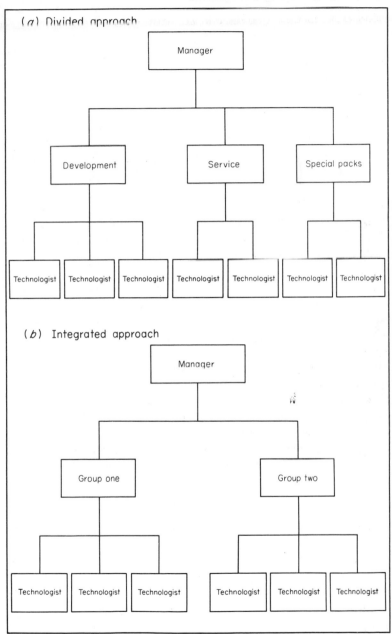

Figure 2.3 Technical packaging organisation

7   Liaison with other company departments.

Close co-operation is always necessary between the technical packaging section and the packaging purchasing unit as areas of responsibility often overlap. In particular, efficient procedures for supplier contact and exchange of information must exist.

Packaging purchasing personnel come from a range of educational disciplines. There is no particular training which serves one well for this function. The main qualities required in a purchasing agent or buyer are that he is commercially sound and that he communicates well with people.

Technical packaging

This section is often the largest in the department, both from the viewpoint of manpower and of facilities. The responsibility for overall co-ordination of the development programme usually rests with this section.

As with the purchasing section it is possible to organise the unit either on a product type or packaging material type approach. The latter method is the most used. This is because to function effectively the packaging technologists need to have a detailed knowledge of all aspects of the materials with which they are dealing. This is more easily achieved if the number of different materials handled is limited. Another possible choice open to the organisation is whether to consider packaging development separately from routine problems or whether to deal with both aspects together. Figure 2.3 illustrates two possible organisation charts for this packaging section.

The first method divides the section into three groups, development, service and special packs. Each group having a number of technologists depending on the work load.

The development group is concerned only with the planning and execution of the packaging development programme. This is usually split between the available technologists on a package type basis. For large individual developments, one technologist may be assigned to handle all technical aspects of the development. In a chemical company, for example, the breakdown of the main type of packaging materials within the development group may be as follows:

1   Unit 1: Metal containers, drums, kegs, cans, aerosols.
2   Unit 2: Sacks — paper, plastic, composite.

3 Unit 3: Bulk and semi-bulk containers.

The service group is responsible for the packaging of existing products and deals with technical problems and complaints concerning existing packaging materials. It usually also has overall responsibility for packaging specifications, standards and test methods.

A third group, responsible for special packs, is also included, especially for retail consumer product type companies. This group is responsible for the development of special promotional packs for existing products. Full details of the types of packs handled by this section are given in Chapter Ten ('Packaging as a Marketing Tool'). The work load in this group can be split between unit containers and outer containers.

The main responsibilities of the technical packaging section can be summarised as follows:

1 Overall responsibility for all technical aspects of packaging.
2 Preparation and maintenance of packaging material specifications.
3 Preparation of the packaging development programme and the setting of objectives.
4 Co-ordination of the packaging development programme. Information supplied to, and communications with, other company departments.
5 Maintenance of quality levels for current packaging material suppliers. Complaint investigation. New supplier approval.
6 Development of promotional or special packs as required by marketing.
7 Technical liaison with suppliers. Continuous review of new packaging materials and methods.

The packaging section is staffed by technologists, but the type of scientific or technical training is not critical. However, the nature of the work tends to attract more chemists than workers from other disciplines. In the USA several universities offer packaging degree courses and hence supply readily trained technologists to industry (see Chapter Thirteen for details of packaging education and training methods).

The art of effective communications is of equal importance to technical competence and plays a big part in the success of this section.

Packaging design

This section has the overall responsibility for the finished pack appearance; both the shape and surface decorations. In consumer goods companies the packaging design section usually acts as a liaison unit between the marketing department and the external designers. They translate the marketing requirements for the pack into the form of a design brief and employ outside designers or agencies to carry out the work. They also ensure that the company image is maintained in individual design projects.

The individuals in the design section are often artistically inclined and can help the marketing people to judge the finished designs. The work of the section is split between the individual members usually on the basis of product type.

In industrial goods companies the design unit is often technically based and is responsible for the actual physical design of the pack. For example, in an electronics company the design group would design the cushioning and outer cases required for the product. Again in this type of situation the design section often delegates the work to the packaging supplier.

The main responsibilities of the section can be summarised as follows:

1 Overall responsibility for package shape and surface decorations.
2 Preparation and administration of budgets for the design programme.
3 Selection of, and liaison with, outside designers or agencies.
4 Preparation of design briefs for individual projects.
5 Setting up and achieving company design objectives. This includes development of company identification by means of a logo.
6 Preparation of standard colour charts and approval of colour standards.
7 Preparation of decorated 'mock-ups' for final pack approval by management.
8 Artwork preparation for finished pack.
9 Liaison with suppliers and other departments.

The work of the packaging design section is obviously closely allied to that of the marketing department, and, in fact, many organisations allocate to the marketing function direct responsibility for package design. The section often also supplies a service to other departments for materials closely allied to packaging, for example display material, coupons, promotional stickers etc. One of the main skills of this section's work lies in evaluating the quality of the work produced by the individual designers or agencies employed.

Packaging engineering

The majority of packaging development programmes involve engineering work concerning packaging equipment. The degree of involvement ranges from the installation of a completely new line, down to minor changes to existing equipment. The packaging engineering section has the responsibility for all aspects of this work.

The purchase of new equipment usually involves the selection of a standard piece of equipment and arranging for minor modifications to meet individual requirements. The modifications may be necessary to handle a special pack design or simply to meet specific safety requirements. In some instances, however, machinery will be custom built to meet a specific new requirement. Many manufacturers now market packaging machinery and materials on a 'systems' basis, ie they supply both the materials and the machinery to handle them. There has also been a rapid increase in the last five years in in-plant conversion of packaging materials, particularly in the fast growing plastic packaging area. Changes to current equipment may involve physical modifications to the machinery or simply the purchase of change parts from the original supplier. Changes in machine operating conditions such as line speed, pressure, temperature, etc are often encountered.

The usual method of organising this section is to allocate the work between the engineers available, on a project basis. The leader of the section is responsible for the allocation of work and also for co-ordinating the projects in hand.

The main responsibilities can be summarised as follows:

1 Recommendation of new packaging equipment either for new projects or as a replacement for existing equipment.
2 Preparation of cost justifications for new equipment purchases.
3 Specification, purchase and approval of new equipment.
4 Installation and start-up of new machinery.
5 Preparation of annual packaging equipment budgets.
6 Investigation of major problems with existing packaging lines.

7 Up-to-date study of new packaging equipment and systems.

The section is usually staffed by mechanical engineers. A good basic knowledge of packaging materials is also required. During the development programmes close contact is maintained with the technical packaging section to ensure that no problems are encountered at the machinery/materials interface.

The organisation outlined in the above section does not exist in total in a large number of companies. The individual components discussed, or composites of them are, however, common. The reader after studying this section can decide how far the responsibility of the packaging department should extend within his own individual operation. For example many packaging departments consist of a technical packaging section and a packaging design section with the responsibilities for purchasing and engineering within the main company departments. A study of the breakdown of responsibilities outlined in this chapter should also help the packaging supplier to understand the allocation of responsibilities within the companies with which he is dealing.

Corporate packaging organisations

Large companies with operations in many product areas, or in a wide range of geographical locations, often have an organisation at their head office to guide the overall packaging operations. Packaging must obviously play an important part in the company's overall structure to justify the employment of this type of unit. These departments are found especially in the large international organisations, such as Unilever, ICI, Shell, BASF, Procter and Gamble, Avon and General Foods.

The corporate packaging organisation is staffed by packaging professionals who have had practical experience with all types of machinery and materials. Often prior to joining the head office unit, they will have had packaging responsibility at one of the individual company locations. Generally the groups are not large, consisting usually of a corporate packaging director, or manager, and a team of packaging advisers. As with the packaging department organisations, the responsibilities can be split in two main ways, either on a geographical basis or on a product group approach. Figure 2.4 shows typical organisation charts illustrating these two approaches.

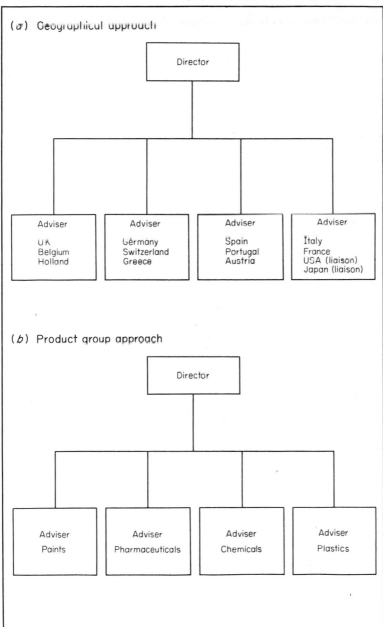

(a) Geographical approach

Director

Adviser

UK
Belgium
Holland

Adviser

Germany
Switzerland
Greece

Adviser

Spain
Portugal
Austria

Adviser

Italy
France
USA (liaison)
Japan (liaison)

(b) Product group approach

Director

Adviser
Paints

Adviser
Pharmaceuticals

Adviser
Chemicals

Adviser
Plastics

Figure 2.4 Corporate packaging organisation

31

Chart '(a)' represents a company with a head office in London and locations throughout Europe. The countries are allocated to each adviser on the basis of estimated work load rather than on geographical position. One of the advisers is also responsible for liaison with associate companies in the USA and Japan.

The product group approach is shown in the second chart in Figure 2.4. Each adviser is responsible for one product area regardless of location. For example, responsibility for the packaging of paint would involve dealing with marketing and manufacturing operations in several countries.

The main responsibilities of the corporate packaging department are as follows.

Planning

Centralised manufacturing organisations are often used to supply packed products to several areas or countries. Each individual area or country supplied will have its own packaging requirements, especially in terms of surface design. One of the responsibilities of the packaging adviser is to liaise between the individual marketing operations and the manufacturing units to ensure that packaging requirements are met. This can often be a very difficult area because of the basic clash of objectives. The manufacturing operations will be trying to maximise efficiency by maintaining the lowest possible packaging material inventories; the marketing organisations in individual companies will often have unique packaging requirements. The packaging adviser has to find a compromise solution acceptable to both.

Project management

Companies using the corporate packaging type of operation usually have people at the individual operating locations with specific packaging responsibility. In manufacturing operations for example, within the engineering departments, a group may be responsible for packaging. When major capital projects are commissioned involving new packaging operations, the packaging adviser is often called in to take overall control of the project. He will manage the project and arrange for the detailed work to be carried out by the regular packaging staff.

Standards and procedures

The head office unit has overall responsibility for the packaging standards and procedures used within the company. The approach to packaging specifications is set by them and administrative details such as format, numbering systems, distribution etc, are outlined. Procedures to be adopted in packaging work such as development methods, budgeting, reporting of complaints, etc, are also laid down. Details of such standards and procedures are usually included in a company packaging manual. The packaging advisers carry out periodic audits to confirm that individual units are complying with the instructions laid down in this manual.

Technical advice

Supply of advice to operating units on packaging matters is one of the main functions of this section. The unit acts as a clearing house for all the packaging activities within the company. It can recommend improved packaging methods to individual areas based on experience in other parts of the company. Equally important it can use this experience to prevent mistakes occurring in other areas.

Personnel

The corporate packaging section has a staff responsibility for all the people in the company with specific packaging jobs. The section is often involved in the selection and training of new packaging personnel. Training of current employees by internal or external courses is also arranged. Often head office packaging courses are organised by the unit for the training of personnel.

Committee representation

The company is represented by members of the section on the national and international committees involved with packaging legislation or standards. Typical of these are:

1 IMCO (Inter-governmental Maritime Consultative Organisation).
2 IATA (International Air Transport Association).
3 BSI (British Standards Institution).

Members of the section are also often active in their packaging associations, for example, the Institute of Packaging (UK), Packaging Institute (USA), Packaging Association of Canada, and the Netherlands Verpackingscentrum (Holland).

New developments

One of the most important responsibilities of this section is to monitor new developments in all areas of packaging. There are three main methods of achieving this:

1 By close contact with all major packaging suppliers; often, large companies work jointly with their suppliers to develop new packaging materials, methods or systems.
2 By attendance at packaging exhibitions and conferences. These offer the opportunity to talk to experts in many different fields and to see new materials and equipment. The leading packaging exhibitions which summarise the trends in both materials and machines are: Interpack (Düsseldorf, Germany), Pakex (London, England); PMM1 (USA – variable location) and the AMA show (USA – also variable location).
3 By obtaining information on the latest developments in the packaging press.

All three methods are used by the team of packaging advisers in their search for new methods which may be applicable to their own operations.

Packaging committee/packaging coordinator

The considerations described above applied to the large corporate type of organisation. The following deals with the medium and small sized companies where packaging does not justify a departmental approach. In the smaller companies packaging may play as vital a part as in the larger ones in the success or failure of the products. The scope of packaging activity, however, is limited either by the size of the annual packaging material budget or by the limited range of packaging materials used. The use of a packaging committee or a packaging coordinator are the two main methods used to manage the packaging programme in these situations.

Packaging committee

The committee method was one of the earliest organisational ways of handling the packaging functions. With the growth of packaging, both in terms of volume and complexity, this method has become less popular. It is still used, however, in one form or another in a large number of companies. In the USA for example, it is estimated that about one third of the major packaging material users still employ a packaging committee form of organisation.

The packaging committee usually consists of representatives of the areas most concerned with packaging. A typical list of members would be:

1 Marketing
2 Manufacturing
3 Engineering
4 Research
5 Purchasing

The packaging representative would also be present, if existing as a separate entity. Departments with a minor interest in packaging such as legal and finance may be called in for specific meetings for example, legal for final pack copy approval; finance for project cost evaluations. A good chairman is a pre-requisite for this type of committee as invariably the individual members have differing attitudes to packaging.

The main advantage of the packaging committee is that it ensures effective communications between all the departments concerned. It also draws upon the total resources of the company to solve the problems encountered in the packaging area, and to progress new developments as rapidly as possible.

Unfortunately, in most situations, the advantages are outweighed by the drawbacks of the system. The main one is that packaging tends to come low on the priority list of the individual committee members, in many companies. The purchasing manager for example, may be much more concerned with basic commodity purchases whilst the marketing manager may be occupied with a new promotional programme. This makes convening of the meetings difficult and often a full quorum is not obtained. The second disadvantage is that it is difficult to keep up to date with the new developments in packaging, when no specific area has complete responsibility for this subject.

Packaging coordinator

Companies which move away from the packaging committee approach often decide to appoint a single individual as a packaging coordinator. This type of position has the same responsibilities as the packaging department (see earlier) but on a smaller scale.

The two main assets of a person occupying this type of position are that he is a good communicator and that he has a good knowledge of a wide range of packaging materials and machinery. The ability to communicate well is vital as he will have no direct authority over most of the departments with which he works. He will also have to represent the company in dealings with a wide range of outside suppliers. Part of the communications requirement will be that the co-ordinator prepares the packaging development programmes and budgets, and places the projects in order of priority. The technical requirement is not so critical, because providing he has a basic packaging training, the incumbent can often learn about new packaging materials by actual experience in working with them. In this situation the packaging suppliers can be of great assistance.

One of the latest trends in packaging organisation is the employment of external consultants to carry out the packaging coordinator role on a part time basis. This is usually arranged by retaining the consultant for an annual fee. This method has the advantage that the company receives advice from a packaging expert who is available to assist with problems or new developments as they arise. This is particularly true for the smaller company who cannot justify a separate packaging function but still requires expert advice.

Packaging supplier

This chapter has dealt with the current organisation methods in packaging material user companies and has discussed the trend to the formation of specific packaging functions, either in the form of a

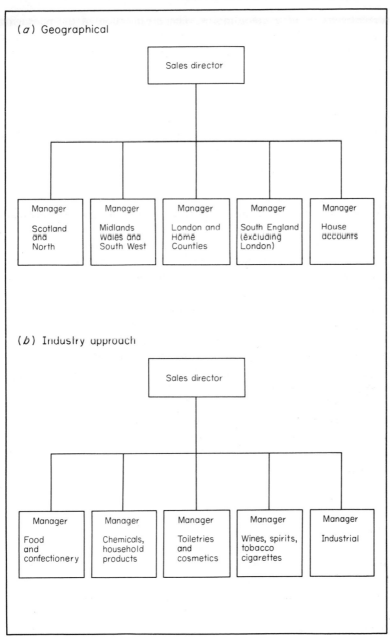

Figure 2.5 Packaging supplier organisation charts

department or of a co-ordinator. The organisation of the packaging supplier has in turn been influenced by the greater attention being paid to packaging in the user companies. In the early days the only contact between the supplier and the user was through the purchasing or buying department. With the growth of separate packaging functions within the user companies there has been an accompanying need for direct technical contact with the suppliers. This has lead to the growth of technical organisations alongside the commercial or sales department within the suppliers organisations.

Sales department

This can be organised in two main ways, either on a geographical basis or by using an industry approach. Figure 2.5 shows organisation charts illustrating the two approaches for a company supplying paper board cartons. The first method, organising on a geographical basis, is the traditional one and is still used by the majority of suppliers today. The country is split into sales regions, handled by area managers, each with a team of salesmen. Often major customers (house accounts) are dealt with directly from the head office by a special accounts manager.

Recently many large suppliers in the UK have switched to the second organisational method, ie using an industry approach. The aim of this is to provide a more specialised service to their customers. Using this method, the individual salesmen are located in the areas most convenient for the majority of their customers.

The sales department is the primary contact with the user company. It has the overall responsibility for ensuring that the customers are satisfied both with the quality of the product supplied and with the back up technical or sales service.

Technical service department

This is usually much smaller than the sales department, consisting of a manager and a team of technicians. They supply technical information to the sales department. They are also responsible for supplying technical assistance to the user company. This will generally involve two main areas, packaging development and packaging service. The development work involves supply of trial material, machinery evalua- tion, pack testing, etc. The service side deals with the investigation of complaints or problems concerning current packaging materials.

This department is responsible for ensuring that all contacts with the supplier are reported to the sales department. The technical services section is often a subsidiary part of the supplier's overall technical department. Many suppliers also have a separate research department for dealing with more fundamental problems and projects.

In this chapter the various methods, currently in use, in approaching packaging organisation have been discussed. In deciding which method is best suited to his own operation, the manager must carefully weigh up his company's own individual requirements. This assessment will include a measure of the importance of packaging to the company. Having done this he can then decide which of the methods outlined will function most efficiently within his own operation.

3

Approach to packaging Development—Planning

Development of new packaging for current or new products is one of the major responsibilities of the packaging department or function in a user company. Chapter Two described the position of the packaging development unit within a packaging department. For the smaller company, having often only a single individual with the packaging responsibility, packaging development work will occupy the main proportion of the work load.

Packaging development can be defined as a project which has as its objective a change in, or an issue of, a new packaging specification. In other words, implementation of a packaging change which will improve the pack performance in relation to one of the five basic packaging criteria (see Chapter One).

Packaging development work can be considered in two separate stages; planning and execution. This chapter will deal with the planning process. Chapter Four will deal with the execution part of the development.

Packaging development projects can vary in size from a small change in specification (caused by a product weight change), to a completely new product involving many new pack components. Effective planning for a packaging development project can eliminate many potential problems which might otherwise be encountered in the course of the development. This applies both to technical problems and communication difficulties with other departments. Five basic planning steps will be considered:

1 Identifying development type
2 Setting objectives
3 Information collection
4 Preparing time-table
5 Cost estimate

These stages will now be considered in detail.

Identifying development type

The first step to recognise in planning the development is the type and complexity of the project in hand. The possible types of packaging development projects in terms of changes in product and pack are:

1 New pack/New product
2 Major pack change/Current product
3 Major pack change/Major product change
4 Minor pack change/Minor product change
5 Minor pack change/Current product
6 Promotional pack/Current product

The different type of projects and the distinction between major and minor pack changes are best illustrated by some typical examples.

New pack/New product

Major developments where both product and package are new are covered here. Examples of this type are:

1 A new convenience food product packed in a vacuum packed flexible pouch.
2 A men's toiletry product packed in a custom blow-moulded PVC or polyethylene container.

3   An expanded polystyrene pack for a cassette tape-recorder.
4   An aerosol all-purpose cleaner.

With this type of project close contact between product and packaging development functions is vital. Often the two project areas are combined on the same critical path time-table.

Major pack change/Current product

Major pack changes which are introduced to improve the package appearance or function, or to reduce the overall pack cost are included in this section. No changes are made to the product itself. Typical examples are:

1   Replacement of tinplate containers (for liquid detergents or lubricating oil), by blow-moulded polyethylene containers.
2   Development of a shrink-wrapped display type outer case to replace a corrugated outer case.
3   Development of thermo-formed PVC containers to replace more expensive injection-moulded ABS-containers for soft margarine.
4   A change from paper to plastic sacks for fertiliser packaging.
5   An improved valve system for an aerosol container.

The majority of packaging development projects fall into this category. In some instances developments of this type can improve the pack in more than one area. For example, improvements in package appearance or function can often be achieved using new packaging techniques and at the same time affecting an overall cost reduction. The first example given above of the replacement of tinplate containers by blow moulded polyethylene containers is a good illustration of this occurring.

Major pack change/Major product change

Major changes in product form which result in the specification of a different type of package are covered under this heading. Examples are:

1   Development of an aerosol personal deodorant to replace a roll-on or stick type product.
2   Change from a powder to a liquid formulation for a household cleaner.

3 Change from a solid to a liquid formulation for a shoe polish.
4 A solvent change for a liquid insecticide formulation involving a change from plastic containers to tinplate containers.
5 Introduction of freeze-dried instant coffee to replace powdered or granulated type of product requires inert gas flushing on packaging line.

Again in this type of development, continued close contact is necessary between the product development and packaging functions.

Minor pack change/Minor product change

Changes in product form do not appear very often. The main type of product change encountered is one of an alteration in product formula, usually retaining the same form. These changes are made to improve product performance or perhaps to reduce overall costs. Even minor changes in product formulation can lead to the need to alter packaging specifications. Examples of this type are as follows:

1 Change in liquid shampoo formulation requires new polymer grade specifications for polyethylene blow-moulded containers to prevent environmental stress cracking.
2 Change in product design for an electrical appliance requires modification to the outer case and cushioning pieces design.
3 Change in powder detergent formulation requires a carton with greater resistance to moisture vapour transmission to prevent caking.
4 Change in recipe for a biscuit type product requires a more protective package to prevent product breakage.
5 Change from a solvent based to a water based formulation for an aerosol product requires a change in valve and can specification.

All product formulation changes should be screened by the packaging function and the effect on pack performance or compatability assessed. If any doubt exists a packaging development programme should be initiated to recheck the suitability of the pack.

Minor pack change/Current product

Minor changes to packaging specifications to reduce costs, to improve the pack function or perhaps to improve manufacturing efficiency are included in this section. Changes in pack size can also be considered here.

1 Reduction of the gauge of aluminium foil used in a flexible laminate.
2 Specification of a lower container weight for a steel drum used to pack a chemical product.
3 New dispensing type closures for plastic containers.
4 Improved valve system for an aerosol container.
5 Change from Imperial to Metric units for a pack range eg 16 oz., 32 oz.and 1 gallon packs become ½ litre, 1 litre and 5 litres.

Promotional pack/Current product

Development of special packs for a given product to meet a promotional requirement is a fast growing area in today's supermarket/hypermarket environments. Typical examples are:

1 Larger aerosol containers to promote 'extra 2oz free'.
2 Packaging of instant coffee in a 'free storage jar'.
3 'Two for the price of one' banded packs for powdered dessert products.
4 Special shrink-wrapped display outers for cash and carry outlets.

Some larger consumer goods companies have a separate section within their packaging department purely to work in this area. They supply special packs to meet marketing requirements.

Setting objectives

Setting objectives for the particular development is the next stage in the planning process. Ideas for new packaging developments will come from any of the areas concerned with packaging. Often the suggestions are made to fit an individual department's requirements. Examples of this are:

1 Marketing require improved pack appearance to compete more effectively with competition.

Company name
Packaging department

Project No.   **Title:**   **Date:**

1 Priority:

2 Background:

3 Objective:

4 Pack description:

5 Work plan:

6 Costs:

7 Timing:

Circulation list   Approvals _____

                                                        _____

                                                        _____

Figure 3.1 Packaging objective sheet

2 Purchasing require lower costs.
3 Production require improved efficiency and line speeds.
4 Engineering wish to use better, faster machinery.

Many developments will also stem directly from the packaging department itself. It is one of the responsibilities of this department to work continuously both with other departments within the company and with outside suppliers, to stimulate new development ideas. The particular requirements for a new development should be drawn together by the packaging department and the objective set in terms of timing and cost. Figure 3.1 illustrates a form layout for this purpose.

The headings give details of the project number, title of the development and the date of origination. Titles should be as concise as possible. For example, the title for the development of a plastic container for a hair shampoo product would simply state the product brand name and the words plastic container.

The background section gives brief details of the reasons for initiating the development and also of the current packaging methods. For new products, mention would also be made here of the basic product concept and of the marketing objective.

The objective restates the title in more concrete terms. For example in the hair shampoo example the objective would state 'to develop a blow–moulded PVC container for the brand name shampoo'. Under pack description details are given of the proposed packaging materials and of the unit and multiple pack sizes to be developed.

The work plan section gives details of the main steps in the development and the allocated areas of responsibility. Reference can be made here to the critical path time-table for the development.

The cost section gives details of estimates of the complete costs to be incurred in the development. These will include laboratory work, development costs, consumer research, package design, machinery charges and the cost of art works and other originations. Details are also given of estimated cost savings in cost reduction projects. Under timing the completion date is stated together with details of the main action dates. The form is completed by an approvals section and a circulation list.

In any large organisation the main packaging development objectives are drawn up twelve to eighteen months in advance to tie in with new product developments and marketing plans. This method has an

---

Company name

Product development department

Product information sheet : brand name detergent

---

1 PHYSICAL FORM: Agglomerated spray dried powder

2 CHEMICAL
  FORMULA:
Full details supplied.
Main components
Alkyl benzene sulphonate
Tripolyphosphate
Perborate or Enzyme
Carboxy-methyl cellulose
Sodium Sulphate
Soldium Silicate
Moisture
Optical Brighteners
Perfume

3 PHYSICAL
  PROPERTIES:
Equilibrium Relative * (ERH)
30% at 68ºF.

4 CHEMICAL
  PROPERTIES:
Deliquescent under most storage
conditions. Product tends to
solidify (cake) at moisture
content of greater than 7%.
Fugitive perfume ie protection
against evaporation required.

5 PRODUCT COST: £ x / ton

6 SAFETY/TOXICITY
  HAZARDS:
None: follow normal detergent
usage conditions.

* ERH = Humidity condition at which product is at
equilibrium, ie neither gives up or takes on moisture.

Figure 3.2 Product information sheet

advantage to the packaging department in that it gives a measure of anticipated work load. This allows for forward planning of personnel, laboratory space and equipment.

Packaging objectives, which normally appear on a single sheet may also be assigned priorities in relation to their relative degree of importance. Packaging objectives often cannot be finally completed until the development planning itself is complete.

Information collection

Before starting a new packaging development the packaging engineer or chemist needs to consider the possible effects of the change or new product introduction on all the areas concerned with packaging. He also needs to be aware of restrictions placed on him by regulations or other factors within these departments. The next step, therefore, in the planning process, is to carry out a systematic check to obtain all the information necessary to proceed with the development.

The following gives a check list of the points which should be covered.

Product

1 Physical form: solid, liquid or other.
2 Physical properties: melting point, boiling point, viscosity, vapour pressure etc.
3 Chemical formulation: chemical name and formula.
4 Chemical properties: for example; corrosive, deliquescent, efflorescent, heat sensitive, light sensitive, oxygen sensitive, etc.
5 General characteristics: abrasive, fragile, agglomerated etc.
6 Names of similar products.
7 Cost per unit of product: £x per ton.
8 Safety hazards: flashpoint, effect on skin, toxicity or other.

Experience in a given product field will lead to a knowledge of the critical items of product information which will affect the particular development. Often standard sheets are provided for the supply of such product information. Figure 3.2 gives an example of a Product Information Sheet supplied to the packaging function giving properties of a powdered detergent product. For this type of product for UK distribution, adoption of a paper board carton employing barrier protection by means of either polyethylene or wax lamination, was necessary to give the required product protection.

Marketing

The basic marketing data which is required, can be obtained from the marketing objectives which lay down most of the information relevant to the packaging. The main points are as follows:

1 Brand information: name, image and basic design.
2 Pack sizes: unit size and shipping outer case sizes.
3 Additional information: coupons, leaflets, stickers etc.
4 Display details: types and sizes of display containers required. Method of in store display eg shelf, dump bins, end aisle, etc.
5 Shelf life: required shelf life before purchase and after purchase.
6 Outlets: type of sales outlet eg supermarkets, self-service, cash and carry etc. Percentage breakdown of the markets by outlet types.
7 Special requirements: for example pilfer-proof, tamper-proof ease of opening, ease of dispensing etc.
8 Shipment areas: geographical selling areas.
9 Future promotional packs: multi-packs, special packs, introductory offers etc.
10 Competitive products: type of packaging employed for competitive products.
11 Cost targets:
   (a) Unit pack costs.
   (b) Development costs.

It will always be difficult to obtain all the above information at the commencement of a development particularly in the case of a new product. The person with the packaging responsibility should press to obtain as much information as possible because any of the factors stated above can influence final packaging choice.

Production

Production requirements usually have a major effect on any packaging development. In the cases of completely new product developments, current manufacturing facilities and packaging equipment are often employed. A study of production requirements should cover:

1 Location: details of plant location and also any details relevant to the location area.
2 Packaging equipment: types available, speeds, capacity, flexibility etc.
3 Custom packers: available facilities with details of machines.
4 Material handling: methods available for storage and handling of empty packaging materials.
5 Warehousing: methods of handling, stacking and storing of packed product.

Distribution

Method of distribution can have a significant effect on both primary and secondary package construction. Changes in distribution systems can lead to the need for a new approach to the packaging of a given product, for example a change to containerised freight shipments often leads to a reduction in the amount of outer packaging necessary. Details which should be covered at this stage of the planning operation are as follows:

1 Loading and unloading: method of transport to and from the shipment vehicle.
2 Shipment method: rail, road, sea, air or any combination of these.
3 Depot locations: situation of depots throughout the distribution area.
4 Climatic conditions: conditions of temperature and humidity encountered during storage and distribution. Possible exposure to water, (rain, sea water or condensation) during distribution.
5 Stacking-heights: details of the method and height of stacking likely to be encountered within the distribution system or at the customer's warehouse.

Consumer

One of the primary considerations is of course the effect of the consumers' requirements on the type of packaging to be developed. Useful information on packaging requirements can be obtained from the basic Consumer Research Life Style or Habit studies. These will give basic information on methods of opening or storage as well as information on likes and dislikes of current packaging. Points which should be covered here are as follows:

1 Method of opening: manual, can opener, knife, tear tape or other.
2 Dispensing method: squeeze, pour, cut or other.
3 Method of storage: cupboard, refrigerator, window ledge, under sink, garage or other. Conditions of storage should be stated wherever possible in terms of temperature and humidity.
4 Frequency of use: number of times product used per week/month/ year.
5 Disposability: method of package disposal. Returnable or non-returnable, re-usable or not. Dustbin or incinerator.

Basic information on consumer habits and attitude with reference to packaging can be built up over a period of time to provide valuable information for future developments.

Legal

The relation between packaging and the law is fully covered in Chapter Eleven. At this stage it is necessary to check under which regulations the product/pack combination falls and obtain details of the relevant information. The following is a list of some of the areas of regulation which can affect package evaluation:

1 Weights and measures
2 Post Office
3 Railways Board
4 Labelling of food regulations
5 Labelling of hazardous goods

This completes the information collection of the planning cycle. This information can now be processed and all the factors affecting the pack listed. From a study of this list the areas of necessary development work can be fully identified.

Timetable preparation

Preparation of a timetable incorporating all the individual activities is the next stage in the planning process. The recommended method of doing this is in the form of a simple bar chart. This can be expanded to a full critical path chart for larger, more complex developments. These methods are a simplified form of PERT (Program Evaluation and Review Technique) and have the advantages of not requiring computer application in the simple form.

This method sets out the tasks to be completed in time order and identifies the period allowed for each activity. It also specifies the responsibility for individual jobs. To illustrate this method, an example of a package development of a minor change in the packaging of a current product will be considered. Consider that a project has been set up to reduce the cost of an aluminium foil laminate for a convenience food product. Three possible cheaper laminate specifications are suggested by interested packaging suppliers. Figure 3.3 shows the bar

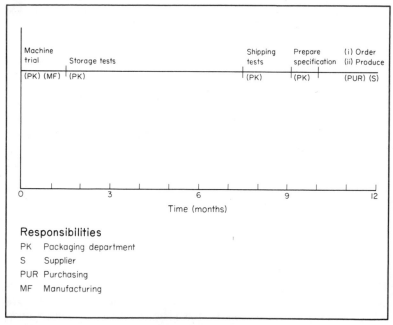

Figure 3.3 Packaging development bar chart

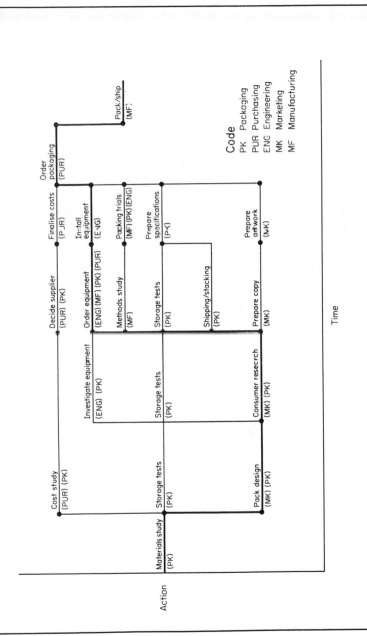

Figure 3.4 Critical path timetable

Code
PK Packaging
PUR Purchasing
ENG Engineering
MK Marketing
MF Manufacturing

53

chart for this development. The development commences with machine trials on the three alternative laminates. Two of the candidates perform successfully on these machine trials and are immediately subjected to storage test evaluation. Following successful storage testing, shipping tests are carried out to confirm the total laminate acceptability. Specifications are then prepared to cover the alternative laminate constructions. Purchasing are now in a position to place orders for the new materials.

In the bar chart the description above the line describes the task and below the line the individual responsibilities are marked. Progress of the development can be marked off on the chart against the bottom of the scale as time progresses. This example illustrates a simple case where all actions in the development are sequential. In a case where several actions are taking place at the same time, the situation is obviously more complicated. In this instance a combination of several bar charts is used and a critical path drawn showing the activities with the longest time period in each section. This method highlights the areas where effort must be concentrated to ensure that the project is completed on time.

Figure 3.4 shows a critical path chart for a completely new product/ new package development. Details are as for the simple bar chart with the job description above the line and the departmental responsibility noted below the line. The critical path can then be drawn in linking the tasks with the longest lead time in each section. In Figure 3.4 the heavy black line depicts the theoretical critical path. This may change during the course of the development.

This type of chart serves the following purposes:

1 Identifies the critical jobs and responsibilities at any given time.
2 Identifies non-critical jobs and allows a measure of the time available for completion. This assists in overall priority rating of the jobs in hand.
3 Provides an up-to-date record of progress on a given packaging development project.

The critical path or bar chart serves as an excellent communication aid. At packaging development progress meetings it gives an immediate picture to all concerned of the current situation. It also acts to identify jobs which are behind schedule and to allow for effort to be concen-

1   **Packaging department**
    (a)   Material study
    (b)   Model/sample preparation
    (c)   Storage tests
    (d)   Pack design
    (e)   Tool/mould costs
    (f)   Test samples
    (g)   Stacking/Transport tests
    (h)   Specification preparation
    (i)   Start-up costs
    (j)   Follow-up costs

2   **Engineering department**
    (a)   Equipment study
    (b)   Equipment specification
    (c)   Equipment testing and approval
    (d)   Installation costs
    (e)   Start-up problems

3   **Purchasing department**
    (a)   Cost study
    (b)   Supplier selection
    (c)   Order production tooling/moulds
    (d)   Initiation costs, printing plates etc
    (e)   Write-off obsolete materials

4   **Marketing department**
    (a)   Pack design costs
    (b)   Consumer research
    (c)   Copy preparation and approval
    (d)   Artwork preparation

5   **Manufacturing department**
    (a)   Methods study
    (b)   Packing/filling trials
    (c)   Start-up costs

Figure 3.5   Packaging development costs

trated in these areas. The drawing up and maintenance of this type of chart takes relatively little time. An experienced packaging person would be expected to prepare a new product/new package development chart in an hour.

The time scales will obviously vary between package type. For example, obtaining samples of a plastic container for evaluation would require purchase of a unit mould taking, say, eight to twelve weeks, whereas sample paperboard cartons can often be obtained within a week.

Cost estimate

Estimating the total cost of the project is the final stage in the planning process. In some cases, particularly those involving complete new product/new package developments, this cannot be achieved until after the first two stages of the development execution (ie the material study and the pack design), are complete. In the majority of situations, however, sufficient information exists at this stage of the planning process to prepare an effective cost estimate.

The activities in the development programme requiring capital investment can be recognised from a study of the critical path timetable. Figure 3.5 gives a breakdown of the possible initiation costs for a packaging development programme. This table can be used as a check list when developing cost estimates. Costs are identified both in terms of capital outlay and labour time. Allocation of overheads such as laboratory equipment will depend on individual company accounting methods.

Packaging department

The main costs incurred in this area are the labour costs involved in carrying out the main parts of the development. Costs are also involved in obtaining sample packages for technical and consumer evaluation. The latter can often be significant, as in the case of blow moulded plastic containers, where the purchase of a unit mould is often necessary to obtain sample containers. Allocation of development costs vary depending on the package type. In the majority of cases the packaging supplier absorbs development costs involved in trial runs for the preparation of samples.

Engineering department

The study, selection and purchase of packaging equipment comprise the main elements of the costs incurred by the engineering department. Details of the methods of machine evaluation are given in Chapter Four. Purchase of change parts for current equipment is also included under this heading.

Purchasing department

Origination costs, whether it be for moulds, for plastic containers or gravure cylinders for printed flexible packaging, are the main items of expenditure in this section. Most areas of printed packaging involve once-off origination charges to cover the cost of plates, screens, or cylinders. Costs for moulds for plastic containers can either be paid in a lump sum at the initiation of the project or can be spread over (amortised) the package price over a given period of time.

Marketing department

Costs incurred for package design and consumer research are two major areas of any new packaging development programme. Minor costs are also incurred in the preparation and approval of copy and art work for the printed packages.

Manufacturing department

The main costs incurred in this area are those associated with trial runs to evaluate packaging materials and machinery. Extra costs are also usually involved due to lower outputs during the start-up period.

In the above, cost areas have been clearly defined within the individual department responsibilities. The actual responsibility for these various cost areas will vary from company to company. For example in some companies the packaging department may have the responsibility for package design and consumer research work. In others this would be the responsibility of the marketing department. Completion of the cost estimate finishes the planning cycle. A careful studied approach to the planning process will ensure that problems encountered during the execution stage (Chapter Four) are kept to the minimum.

4

Approach to Packaging Development—Execution

Chapter Three dealt with the planning stage of the development process and explained that the degree of planning necessary depends on the complexity of the project in hand. The main aim of the planning process is to smooth the progress of the execution part of the development and to ensure that it proceeds as efficiently and economically as possible. The execution, or carrying out, of the development can be considered in five distinct stages, namely:

1 Creative
2 Consumer evaluation
3 Technical evaluation
4 Implementation
5 Start-up

These sections will now be covered in detail.

Creative

In today's competitive supermarket environment differences in pack

Group	Classification	Types
1 Metal	(a) Materials	Steel (plain or lacquered), aluminium, tinplate (plain or lacquered)
	(b) Containers	Drums, kegs, cans, aerosols, pails, tubes
2 Plastic	(a) Materials	Polyethylene (low and high density), PVC, polyvinylidene chloride, polystyrene, polypropylene, polyester, ABS, Phenol formaldehyde, urea formaldehyde, polycarbonate, polyacetate
	(b) Containers	Blow moulded, injection moulded, compression moulded, film overwrap or shrinkwrap, thermoformed tubs, trays, etc, bags, sacks, tubes
3 Paper	(a) Materials	Paper, board; plain, coated or laminated, corrugated board; plain, waxed, coated, laminated
	(b) Containers	Labels, cartons, outercases, moulded pulp, sacks, drums, cans, envelopes
4 Wood	(a) Materials	Wood, cellulose film; coated, uncoated or laminated
	(b) Containers	Casks, kegs, crates, barrels, film overwraps, pouches, bags
5 Glass	(a) Materials	Glass; coated or uncoated, decorated or metallised
	(b) Containers	Bottles, jars, carboys, ampoules, tubes
6 Laminates	(a) Materials	Paper/plastic film, aluminium foil/paper, aluminium foil/plastic film, plastic film/plastic film, plastic film/cellulose film, cellulose film, multiple combinations
	(b) Containers	Overwraps, pouches, bags, cups
7 Composite	(a) Materials	Paper/plastic, metal/plastic, Paper/plastic/metal, metal/wood/plastic
	(b) Containers	Drums, cans, sacks, kegs

Figure 4.1 Major packaging materials and containers

design, shape or concept, can be significant in accounting for the relative success or failure of all products. This is particularly true in the case of the 'me-too' type of product. Packaging is often the most important item of the total product characteristics. Changes in pack design can be used to re-launch products successfully without any lation change. Typical of this form of development is the introduction of an improved design or presentation for a cigarette product by changes involving packaging surface design or packaging material components.

This section is probably the most vital area of the whole development, particularly in the area of consumer products. In many instances the final product's success or failure rests on the work carried out at the creative stage.

Materials study

The first action on the critical path chart (see Figure 3.4) under the packaging responsibility is the materials study. This involves looking at both the product characteristics, as defined in the planning stage (Chapter Three), and at the full range of available packaging materials. Based on experience of both these sections a list of likely candidate materials can be drawn up.

Figure 4.1 gives a breakdown of the major types of packaging materials and containers which are considered here. This list can be used as a check list for this stage of the creative process. Many variations are obviously possible within each category. In the case of laminates, for example, the number of material combinations is enormous. Luckily a large percentage of these combinations are ruled out on cost and performance considerations. An initial check will give a provisional list, which will be extremely helpful at the design stage.

At this point the likely package constructions selected need evaluation for product compatibility or product protection requirements. Storage tests should be set up and all the possible alternatives should be considered. It is important that these tests are set up as soon as possible so as to provide back-up information at the time when the choice of the final pack is made. Details of storage test methods will be discussed later in this chapter.

Package design

Design of both the physical shape of the package and of the surface decoration is the next activity to be tackled. Design projects can be handled in one of three ways.

Staff designer or department   The user company employs a staff designer, or in larger companies often a design department, to tackle all the necessary package design projects. This approach applies mainly in companies where the accent is more on physical form than on appeal through surface decoration. For example, a designer in a company manufacturing electrical goods would be more concerned with the physical aspects of cushioning and outer case design, than with the finished surface appearance. Larger consumer goods organisations often have a design department whose responsibility it is to liaise with outside designers (see Chapter Two — 'Organisation of the Packaging Function').

Design consultants   Design consultants, or design firms, can be engaged to carry out projects either on an annual retainer basis or on a flat charge per project. This is the route favoured by most firms who have to compete in the supermarket and self-service type of outlet. Companies using agencies on a freelance basis have the advantage that they can vary assignments between different design agencies and hence obtain the maximum amount of creativity. When using a design firm for a given project the services of a design team are obtained who have experience in many fields and can often bring completely new approaches to the project.

Supplier designer   Most major packaging suppliers employ a design department which provides a customer service. No direct charges are made for this type of service but it should be recognised that the overheads for the service will be included in the finished unit packaging material cost. Supplier designers have the advantage that they know well the materials and processes with which they are working. Any design produced by such a designer will almost certainly be practical. They usually, however, suffer from the disadvantage that their creativity is limited by having to produce designs which come within the range of the suppliers' own materials and processes.

It is essential, whichever of the above three methods are followed, that the designer is given as much information as possible concerning the project in hand. This is best done in the form of a written design brief. Figure 4.2 shows a typical form layout for such a design brief.

This form gives several headings which provide the information to the designer. It will not always be possible to provide information in all sections initially. For example, for a completely new product/new pack combination details of the proposed manufacturing method will probably not be known at this stage.

The form commences in a similar fashion to the basic objective form (see Figure 3.1) with a section of background information followed by a statement of the objective.

Under 'Scope', the expected pack sizes, the sales areas, and a description of the methods of handling and transportation, are covered.

Product factors which may affect the packaging choice are mentioned under the 'Protection/Compatibility' heading. A list of materials and containers emerging from the material selection process is also stated here.

Expected 'Usage' method is stated if possible, in quantitative terms. Details of the prospective consumer groups at which the product will be directed are also useful.

Details of the proposed method of packing or filling are included under the heading 'Manufacturing Method'. Restrictions placed on the design by unavoidable manufacturing limitations are mentioned here.

Under 'Visual Criteria', any components which should be incorporated into the design, for example the company logotype, are stated. Also any details of basic consumer objectives are included here, for example 'pack should be easy to handle'.

Under 'Technical Performance', performance requirements which are necessary to meet distribution or usage needs are stated. For example, if a plastic container was being considered, indications of expected impact resistance would be given.

Under the 'Cost' section, details of target costs for the pack, together with maximum limits, are quoted.

Completion dates for the total project, and for this stage of the design exercise, are stated under the 'Timing' heading.

The form is completed by an approvals section and a circulation list.

The actual design project is usually tackled by the designer in two distinct stages:

Company name

Design Brief: Project Title

Background

Objective

Design Information

1 Scope
2 Protection/compatability
3 Usage
4 Manufacturing method
5 Visual criteria
6 Technical performance
7 Cost
8 Timing

Circulation   Approvals _____

                               _____

                               _____

Figure 4.2 Packaging design brief

1 The concept or ideas stage
2 The development stage

The first stage consists of drawing together as many ideas of pack combinations as possible. This applies whether the design project concerns physical aspects, surface decoration or a combination of the two. Brainstorming sessions and other methods of group creativity may be used to stimulate the ideas flow. This stage of the activity is completed by the designer preparing a series of sketches covering the most promising ideas for presentation to the client.

The preliminary sketches are discussed by the user company and the designer, and a few are selected for further development. The designer then prepares scale drawings and mock-ups of the selected designs. The final pack selection can be made following inspection of these mock-ups or a number can be selected for consumer evaluation. It is the responsibility of the user company at this stage to confirm the technical feasibility of the designer's suggestions.

Consumer evaluation

Evaluation of consumer attitudes to pack changes or to new product/new pack combinations, developed by the designer, is the next action to be considered.

For companies concerned with industrial-type markets the evaluation of customer reaction is a relatively simple matter. Trial shipments of the new pack can be sent to major customers for their reaction and comments.

In the field of consumer products the evaluation is more complicated. Changes in either surface design or shape can have a significant effect on consumer acceptance. The problem is to try and obtain a measure of the eye-catching appeal of the pack in the supermarket environment as well as the acceptance by the consumer of the individually viewed pack, in use.

Attitudes to the necessity for consumer research on the pack, vary widely from company to company and even where the necessity is admitted, there are widely differing schools of thought as to the amount that should be carried out. Some companies rely on the designer to produce a pack combination which will have good consumer acceptance. Many designers incorporate consumer research as part of their overall design programme. Other companies believe in carrying out their own detailed testing of all aspects of the pack. The following are the main test methods used.

Informal – house panels

A panel of housewives from the companies' employees are given samples of the proposed new product/new pack combination and asked to use the product at home for a period of two weeks. At the end of this period they are asked to fill in a questionnaire to give a measure of their attitudes to the test samples. The number of people per panel is not critical as these tests are usually used only to get a general idea of the concept acceptance and a measure of any major negatives. No attempt is made to assess the results statistically. This method is useful at the initial stages of a development to screen ideas, or for projects where more extensive research cannot be justified for reasons of cost or timing.

Housewife panel tests

Consumer research panels are run at periodic intervals by market research and consumer goods companies. The normal procedure is to hire a local community hall or meeting centre and invite housewives from the area to attend at a given time. On arrival, the housewives are asked to fill in a questionnaire giving details of their usual brand usage and relevant demographic information. They are then invited to carry out a series of tests covering several different product areas.

Two types of packaging test are used at these panels.

Shelf impact assessment   Two methods are used to attempt to measure the effectiveness of a surface design in the supermarket type environment. The first method consists of setting up a supermarket shelf and placing the test pack in the midst of a range of competitive products. The housewife is asked to walk past the shelf display and afterwards she is asked to list the names of products she has seen on display. The percentage recall of the test pack in relation to the current products gives a measure of shelf impact. During this type of test, the position of the test pack is varied to eliminate any order effect.

The other method consists of a mechanical test, such as the tachisto-scope. A series of pack designs are shown to the housewife through a viewer and she is questioned at the end for pack recall. This method can also be used for a single design to assess the relative legibility/ recognition of key elements of the design, for example brand name, major copy, illustrations, etc.

Look/handle test   Preference between two or more designs is obtained by this type of test. The housewife is either shown, or asked to examine, a set of models representing the test pack. She is then asked to comment on each individual container against a series of phrases. Figure 4.3 shows a questionnaire which was used to compare three prospective designs for a liquid insecticide container. The housewife was told that a new container was being considered for this product and she was asked to look at and handle three prospective shapes, represented by models A, B and C. She was then asked to fill in the questionnaire illustrated in Figure 4.3 for each model and requested to circle the appropriate number in each case. At the end of the test the housewife was also asked to make a choice as to her preference between the three designs. A statistical analysis of the total results gives a measure of acceptance between the three designs.

Placement test

A more expensive method of package evaluation is an actual product placement test. In this method housewives are given two packs each containing identical product formulations. Pack A represents the current pack and Pack B the new trial pack design. The packs may differ in physical appearance as well as in surface design. The housewife is asked to use each product in turn, for example product A for one week, product B for one week. At the end of the two week period she is asked direct questions on her preferences between product A and product B. A typical result of this form of test, where A and B differ only in surface design would be:

> Prefer product A – 22%
> Prefer product B – 13%
> No preference   – 65%

These results indicate that the design of the product A pack is preferred

Code	Look/handle test: Liquid insecticide container						
					A or B or C		
		Agree Strongly	Agree	Neither/ Nor	Disagree	Disagree Strongly	Dont Know
1 This is a modern container		1	2	3	4	5	6
2 I find this container easy to hold and handle		1	2	3	4	5	6
3 This container is easy to pour from		1	2	3	4	5	6
4 This container is attractive to look at		1	2	3	4	5	6
5 This looks like the kind of product I would use at home		1	2	3	4	5	6
6 Looks appropriate for an insecticide		1	2	3	4	5	6
7 Looks like a quality product		1	2	3	4	5	6

Figure 4.3 Consumer research questionnaire

to that of the product B pack. The high degree of 'No preference' result is not unexpected as in this type of test, the product properties account for the main reasons for preference. Results can be analysed statistically to check on the significance of the figures. Housewives are also asked at the end of the test a direct question on the preference between the two pack designs. A typical breakdown of results of answers to this question would be:

> Prefer product A   – 60%
> Prefer product B   – 30%
> No preference   – 10%

In this case where the question is directed specifically at the pack appearance, the number of 'No preference' replies is lower. Providing the number of housewives participating was adequate the result of this type of test would represent a significant preference for the pack design of test product A over that of test product B.

Market tests

Marketing of test packs in individual stores or in test towns or areas is the last common method of consumer pack evaluation. Pack performance in this case is measured in terms of actual sales figures. Many large department stores use this method to evaluate both new pack and new product ideas. They place stock of the new item in a premier position and measure the sales over a period of time. Pack evaluation by in-store, town or area tests is also used by many large consumer goods companies.

These tests are backed up by local paper or TV advertising. The new pack is put on sale and the consumer attitude is measured by either in-store interviews or by door-to-door questioning. This method supplies information from both product users and non-users. It also has the advantage over the other tests that the people interviewed have actually purchased the product and hence hopefully will give an objective appraisal. Information is also obtained as to the pack acceptability by overall figures of sales and market share.

Technical evaluation

Evaluation of the technical aspects of the package begins immediately after the materials study with the setting up of storage tests (see Figure 3.4). The major part of this activity is not, however, completed until results of the consumer evaluation are available.

Storage tests

The two main aims of storage tests are:

1 To confirm product/package compatibility.
2 To assess the shelf life of the packed product.

Product/package compatibility   Interaction between the product and the package which results in a deterioration of either of these components is commonly revealed as a result of storage testing. Some typical examples are as follows:

1 Corrosion of metal containers by liquid insecticide formulations.
2 Environmental stress cracking of blow moulded plastic containers by detergents.
3 Mould or bacterial growth in food packages.
4 Delamination of flexible laminates by chemical products.

When potential hazards of the above nature exist, the usual procedure is to develop an accelerated test which helps to screen changes in either product or package. An accelerated test method to predict whether stress cracking of plastic containers is likely, consists of filling test containers, one-third full of product and storing at 60°C for two days. Containers are examined for signs of stress cracking at the end of the storage period.

Other methods, such as the Bell Stress Crack test, study the effect of detergents on specially moulded test pieces of plastic. It is important that correlation is obtained between this type of short-term accelerated test and actual long-term storage evaluations. This will ensure that acceptable materials are not rejected unnecessarily. In cases where damage during shipment can affect the new product/new package compatibility, shipping tests should be carried out prior to storage testing. An example of this is the evaluation of lacquered drums for the packaging of corrosive chemicals. Damage during transit can cause cracks in the internal lacquer surface of the drums which exposes the metal which may be attacked by the product during storage. Simulated or actual shipping tests are carried out therefore prior to storage testing.

Shelf life determination   Measurement of the shelf life of a packed product is normally carried out in the laboratory under controlled conditions of temperature and humidity. The conditions can be achieved either by the use of cabinets or by specially built storage rooms. Two generally accepted test conditions are:

$$77^{\circ}\text{F} - 75\% \text{ R.H.} - \text{Temperate areas}$$
$$100^{\circ}\text{F} - 90\% \text{ R.H.} - \text{Tropical areas}$$

Many other conditions are used depending on the distribution and shelf life requirements. Accelerated tests are often used to try to get a quick answer to questions concerning either compatibility or shelf life. One method used for moisture sensitive products is to measure the moisture pick-up of test packs for a relatively short time (seven to fourteen days) under accelerated conditions and measure the time taken for the product to reach its critical moisture content. A correction factor is then used to translate this into terms of shelf life under actual usage conditions. For example, in a particular test, a ten day shelf life under tropical conditions may be equivalent to a twenty-one day life at temperate conditions. The main problem with this method is to assess accurately the conversion factor between the accelerated and actual conditions.

Another method commonly used for moisture sensitive products is to store the test packs under normal use conditions and to measure the weight gain with time. As the relation between moisture gain and time is exponential for many packed products, it is relatively easy to extrapolate the results. The logarithmic function of the moisture content is plotted against time and from the straight line graph obtained, the time taken to reach the critical content is measured.

These two methods apply only to moisture sensitive products where the critical moisture content is known. Many products decay on storage without any appreciable gain or loss in weight. With this type of product it is necessary to carry out performance tests at periodic intervals during storage, to measure the shelf life. A typical example of this is a series of tests carried out on flexible laminates for a dried yeast product. Shelf lives for the various laminates were measured by 'baking' tests carried out on these at periodic time intervals during storage.

Wherever possible storage tests under actual use conditions should be carried out. The main danger of accelerated storage tests is that they can lead to over-packaging. Materials which may be adequate for use can be rejected by a too severe accelerated test. There is also the other danger that the accelerated test will not uncover all the defects likely to occur under normal use conditions.

Long term storage tests under actual conditions should measure as much information as possible. A control pack is always included and compared against the test packs for:

1 Visual changes in pack or product.
2 Weight lost or gained.
3 Gas (eg oxygen, nitrogen etc) loss or gain.
4 Formulation changes by analysis.
5 Product performance changes.

Long term storage tests often continue throughout the life of the development. They act as a continuing reassurance that the pack will perform as expected in use. Information from storage test work can be used to improve, or modify packaging specifications, even at advanced stages of the development.

Machine trials

In development programmes where it is desirable to use current packaging equipment the next step is to carry out trials at the manufacturing location to determine the suitability of the new package on the available equipment.

It is important that these trials be treated seriously and that adequate machine time is made available. This is often difficult with production pressures on current items but misleading results can be obtained by carrying out trials with only a few packs for a short period of time. Machine problems undetected at this stage can lead to expensive downtime and production loss at the start-up time. Ideal conditions for a packing or filling trial are as follows.

Product The product used for the trial should be identical to that planned for the finished pack. Slight changes in product properties such as density, flowability, viscosity etc, can have significant effect on the performance of a given packaging machine.

Personnel   All personnel associated with the packing or filling operation should be present for the trial. These will include the production manager or foreman, the operatives and the engineers or fitters responsible for machine maintenance and change-over. Attention should be paid to their views as to the pack's suitability for the given machine operation.

Machine performance   Machine performance under varying situations is evaluated. Conditions such as temperature, dwell time, pressure, etc are assessed in relation to the overall efficiency.

Packaging materials   As with the requirement for product, it is important that the range of packaging materials under test represent fully those which will be used for the finished pack. Full details should be available of the specifications of the packaging materials tested at this time. These materials should if possible be produced on commercial equipment rather than on laboratory or pilot plant machinery.

The trials should be carried out at maximum line speeds and a long enough run obtained to gauge the expected line efficiency. Potential problems and necessary modifications are noted and discussed, where necessary with operating personnel. Changes in pack specification which would improve the machine performance should be looked for and noted.

Warehouse/shipping trials

Chapter Five deals in detail with the physical methods of pack evaluation. In this section the type of tests carried out as a routine part of the technical evaluation are outlined briefly.

Stacking tests   Tests to determine the strength of the package during stacking in warehouses and in transit, can be carried out in two ways:

1 The tests can be simulated in a laboratory using either a static test rig or a dynamic compression tester. These methods give a comparative measure of the resistance to compression and damage during stacking.

2 The more practical method is to set up a stack in a warehouse and observe the effect on the packages for a period of time. Changes in environmental conditions such as changes in relative humidity can have a significant effect on tests of this type.

Stacking tests should always be carried out under the severest conditions likely to be encountered ie maximum conditions of temperature and humidity. If this is not possible then a correction factor should be allowed for in the interpretation of the results, to account for the possibility of more severe conditions being encountered.

Shipping tests   Transport or shipping tests are mainly used to evaluate the suitability of outer cases or other forms of over-packaging. They are also used to confirm that changes in unit pack construction are acceptable.

These tests can either be carried out under simulated conditions in the laboratory or by sending trial packages through typical distribution systems. Chapter Five gives full details of each of these two methods. The simulated laboratory tests, such as the LAB test, drop tests, inclined plane test, etc. give a good comparative measure of likely pack protection. In any packaging development programme actual test shipments should always be carried out to confirm the results of simulated laboratory tests.

Equipment investigation

The purchase of new packaging equipment is often one of the major parts of a packaging development programme. At this stage, and when the pack type and requirements have been decided, the engineer or other person responsible for the purchase, starts to consider the types of equipment available. He will usually have a backlog of information on all the types of machinery currently on the market. This may be in the form of magazine supplements or in files of equipment supplier brochures. Many engineers use the annual or bi-annual packaging exhibitions to keep up-to-date on all the current types of packaging equipment. From this information the engineer draws up a short list of likely machinery.

Meetings are then arranged with the machinery suppliers, at the suppliers' plants, to discuss the machines in more detail. Where possible the machines are inspected actually in operation at another manufacturer's plant. This type of visit is invaluable, for in this way an unbiased opinion of the machine is obtained from the other purchaser. The next step is to rate the machines under a series of headings. Figure 4.4 shows a typical form which can be used for this purpose.

The headings include the machine name and details of the packaging development project to which it applies.

Details of the supplier's name are stated and a brief account of the company's history and trade reputation may also be included under the supplier heading. Any previous experience with machinery from this supplier is also summarised here.

The type of machine and the way in which it will operate to meet the company's requirements are included under 'Description'.

Machine 'Output' and efficiency under the range of operating conditions are included under the next heading. Details of labour requirements for the machine operation are also stated here.

An engineering assessment of the machine covering mechanical, electrical, materials, finish etc is given under the 'Quality' heading. Included in this section is an assessment of the machine's accuracy, for example tolerances on filling weights, etc.

Under 'Cost' the basic machine charge is stated, together with the cost of necessary or optional change parts. The cost of spare parts should also be included here. Details of the method of payment and also of any machine guarantee may also be included here. In cases where the machinery is being purchased to implement a cost saving based project, details are given of the proposed justification.

Under 'Delivery' the time taken for delivery and installation of the machine from the order receipt is given. This time should include an allowance for machine testing and approval at the supplier's plant.

Safety aspects are the next item covered. Assessment of the machine's safety should include both mechanical and electrical aspects. Often suppliers are willing to carry out modifications to meet individual safety requirements.

Details of the complete range of pack types which the machine can handle are given under the 'Flexibility' heading. This can be particularly important where changes in pack size for promotional or other marketing reasons are possible. Changeover times from one pack size to another are also stated here.

The form is completed by an 'Overall Assessment' section which summarises the main good and bad points of the machine.

Company name

**Packaging Machinery Evaluation
Sheet**

Project Name & No.   **Machine Name**

1 Supplier

2 Description

3 Output

4 Quality

5 Cost

6 Delivery

7 Safety

8 Flexibility

9 Overall assessment

Prepared by

Approved by

Distribution

Figure 4.4 Packaging machinery evaluation form   *75*

A form of this type is drawn up for all the machines on the short list. The next step is to compare the reports and prepare a recommendation for the packaging machinery selection. The evaluations can be used as appendices to the report making the final recommendations for packaging machinery purchase.

Methods study

Engineering drawings for the proposed packaging machine are passed on to the manufacturing department following final decisions on packaging machine selection. The plant personnel can now study the proposed method of operation. This study may be carried out by the production manager responsible for the department in which the machine will operate, or in larger companies by a separate industrial engineering section. The main points of the methods study can be broken down as follows.

Layout   Plans are drawn up of the packing line layout and the position of the new machine marked. Often scale models are constructed to assist in the choice of machine location. Figure 4.5 shows a typical packing line layout and illustrates a line for a pouched food product which is packed ten pouches to a paperboard carton. The four form, fill and seal machines at the beginning of the line feed pouches on to a conveyor. The carton erecting machine takes flat carton blanks, forms them into cartons and feeds them on to the conveyor alongside the pouches. The pouches are then placed in the cartons by the female packers who are sitting on both sides of the conveyor. A final packer is located at the end of the conveyor to remove any loose pouches or cartons and also to check that the packed cartons enter the next unit correctly aligned.

The next part of the line is the carton sealing machine. Two flaps are glue sealed in the first operation, the carton is then turned through an angle of 90° and the second pair of flaps are sealed. The cartons then pass on to a packing table where two operators place them twelve to a corrugated outer case. The outer cases are glue sealed and passed under a compression belt after which they are palletised on either side of the packing table.

Labour   The number of personnel required to operate the line is calcu-

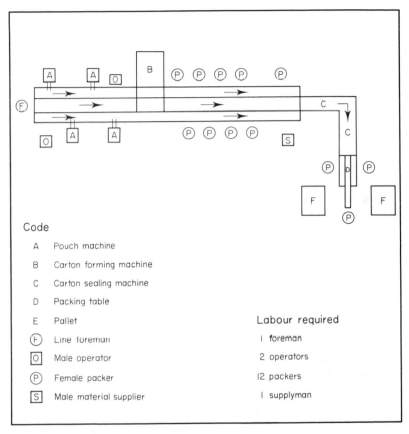

Figure 4.5 Packaging line layout

lated. The positions are marked on the layout drawing as illustrated in Figure 4.5. The degree of competence required for each individual job is also assessed at this stage. The number of people required for each line is determined after a study of machine speeds and predicted efficiency.

Costs   The installation costs of the packing lines are calculated. Often the packaging machinery quotation will include machine installation and the services of an engineer for the start-up period. Extra costs to be assessed at this stage cover the supply of services, eg air, electricity or gas and also the purchase of ancillary equipment such as conveyors and packing tables.

Output   From the information supplied as to the basic machine output, the overall production for a shift operation can be calculated. This is usually measured in terms of outer case units, for example, 320 cases of twenty four units per shift. This figure includes estimates of downtime both for reasons of machine inefficiency and for regular downtime for meal breaks, etc.

Cost study

The final stage of the technical evaluation process is to prepare a costing of the total new product/new pack combination. Most companies have a standard form which states the total costs in terms of a unit pack or multiples of that unit (for.example, 1, 6, 12 or 24).
Costs are broken down as follows.

Product   The cost of product per unit pack, together with an allowance for wastage during packing or filling and distribution are stated.

Packaging materials   Basic packaging material costs are supplied by the purchasing department. At this stage if the suppliers are not finalised the figures used will be estimates based on quotations received to date. An allowance is made for wastage or shrinkage of packaging materials during use. This will vary depending on the package type. For example a flexible laminate for use on a form, fill, seal type machine may have a shrinkage allowance of 5 per cent, whereas for a corrugated outer case, to be packed manually, it will be no more than 1 per cent.

Direct costs   Costs incurred in regular day-to-day operations are included. Labour costs for the packing line personnel and cost for machinery depreciation are the main components here.

Indirect costs   Overhead costs to cover the services supplied are covered in this section. Other ancillary costs such as administrative personnel, research and development etc may also be included in this figure.

Distribution Costs associated with the storage of packed stock in the warehouses and shipment from the factory to depots or retail outlets are stated in this section. Distribution costs usually account for a significant proportion of the total product costs.

The completed cost form is agreed by all the departments concerned and is then incorporated by the accounting or financial department as the official standard cost for the product concerned. It is important when dealing with costs associated with packaging development work to consider all the items stated above. Consideration of basic packaging material costs in isolation can be very misleading.

Implementation

Development of the pack has now reached the stage where the technical and consumer evaluations are complete. Management approval to proceed with the actual implementation is often now necessary. One way of doing this is to draw up an official recommendation outlining the total development project. This can be done in the same form as the Objective Sheet (see Figure 3.1) a 'Recommendation' section replacing the 'Work Plan' paragraph. A copy of the proposed standard cost sheet and a full scale mock-up of the proposed package should accompany this recommendation.

Following approval for the new pack, work commences on the physical development of the pack.

Specification preparation

Specifications are prepared for all components of the proposed pack. These are discussed with all the interested company departments and agreed with the packaging suppliers before approval. Chapter Six deals in detail with the recommended methods for preparation and mainten- ance of the specifications. Following approval of the specifications they are issued to all the departments concerned. The purchasing unit is usually responsible for officially passing on the specifications to the packaging suppliers.

Supplier selection

During the initial stages of the development the purchasing department assists with the cost study and considers potential packaging material suppliers. Initial quotations are then obtained from these suppliers based on provisional specifications supplied by the packaging department. These specifications become available as soon as the consumer research evaluations are complete.

From the approximate quotations received and drawing on previous knowledge of the suppliers, the purchasing agent draws up a short list of potential suppliers. The suppliers on this short list are then requested to quote officially basing their quotes on the approved specifications. For items which the purchasing agent orders regularly, and has extensive experience of the suppliers concerned, the final decision will often be made on the basis of these quotations. Packaging material items such as corrugated outercases, paper labels, paper-board cartons and plastic film are typical examples of the materials ordered on this basis.

In situations where the packaging material is either new to the user company or where significantly large orders are to be placed for the first time, the supplier assessment will be more detailed. An appraisal of each supplier's operation is drawn up by the purchasing agent with the assistance of the technical packaging personnel. Information for this study is obtained by visiting the supplier's plant and inspecting manufacturing, research and commercial facilities. A report is then drawn up on each potential supplier covering the following points.

Commercial service   The purchasing agent makes an assessment of the efficiency of the commercial side of the supplier's operation. This will mainly concern day-to-day contacts with sales representatives and internal sales personnel. He also estimates the reliability of the stated lead times and delivery dates. Lead times are compared with those from alternative suppliers. Potential sources of the supplier's raw material, necessary in the case of an unexpected disruption, for example, by strikes, fires, etc are discussed.

Price   Details of price quotations are compared from the alternative suppliers. Prices are analysed using value analysis techniques to confirm that they are realistic. Future price trends for the particular items involved are predicted.

Quality   The technical packaging personnel measure the supplier's overall quality in two ways:

1 By a visual assessment of the items in production during the plant
   visit. For example, the printing quality of a silk screen printed
   plastic container will be measured by watching printing operations
   and examining stocks, already produced. An in-store check of the
   same container would give a further measure of the printing and
   overall quality.
2 The supplier's method of quality control or quality measurement
   and the facilities available for carrying it out are also studied in
   detail. Test methods in use are discussed in relation to the user's
   specification. The technical competence of the supplier's quality
   control personnel is also rated following these discussions.

Capacity The supplier's production capacity is compared with the
potential order size. Future machinery purchases are discussed with the
supplier if current capacity is insufficient. Many user companies place
annual contracts for packaging material supplies which enables suppliers
to justify the purchase of new plant and equipment.

Financial rating When dealing with new suppliers, particularly the
smaller ones, purchasing agents obtain a report on the supplier's overall
financial conditions. This type of report, supplied by companies such
as Dun & Bradstreet, gives a measure of the likely stability of the
company. It also gives the purchasing agent a measure of whether the
supplier will be a reliable long-term source.

Research and development facilities The supplier's research and development facilities are rated in terms of their potential for assisting the
user company in future development work either on the given project
in hand or with future projects.

   In some situations the above procedure will be carried out and the
supplier's decision taken before packaging material specifications have
been finalised. This is often the case when dealing with items of plastic
packaging material. Moulds for either injection or blow moulding often
have to be ordered to produce samples for evaluation. In these instances,
approximate price quotations are obtained based on the anticipated
design.

   When dealing with large quantities of packaging material two or
more suppliers may be chosen to produce a given item. This is usually

done to ensure a continuing supply position in the case of disruption at one supplier. In these instances one supplier is developed for the initial start-up and second or subsequent suppliers given approved packs from this source, together with specifications, as a target.

Machinery purchase

Three main steps are involved in the purchase of new packaging machinery or complete packaging lines. These follow on from the decisions on the equipment supplier (see previous section).

Purchase order   This gives details of the machine and of engineering, electrical and safety requirements. It should also give explicit details of the machine's performance in use. For example, a purchase order for a powder filling machine would specify allowable weight tolerances at given running conditions. Another example is a glass container packing line which would specify line speeds for individual container sizes.

Machine approval   The purchase order also specifies that approval for shipments must be given by the user following a trial at the supplier's plant. The user arranges to ship the product to the supplier's plant for this trial and conditions should be as near typical of the user's operation as possible. The packaging engineer attends this trial and lists the necessary modifications to be made before the machine is shipped. During this trial the machine performance is measured against the criteria stated in the purchase order. It is also important that the packaging materials used for this machine approval trial conform with the final specifications.

Installation   Machine installation is normally carried out by the supplier's service engineer. As the time spent at the user's plant by this engineer is limited all the necessary preparations are made prior to his arrival. The installation is supervised by the packaging engineer who also tests out the machine to confirm that it meets the requirements as stated in the purchase order. Actual finished packaging materials from the first order are used for this final machine trial. A list of necessary modifications is agreed with the supplier's service engineer. Arrangements are also made with the supplier's service engineer to train machine operators and maintenance engineers in the use of the machine.

Packaging material preparations

Preparation of the material necessary for the production of the finished packaging items is the next action to be carried out at this stage of the development. Great care is necessary to avoid errors during this operation to ensure that the finished package meets the specified appearance requirements.

Layout drawing   A layout drawing is prepared by the packaging department as part of its detailed packaging material specification (see Chapter Six). This illustrates the areas where printing or decoration are to be

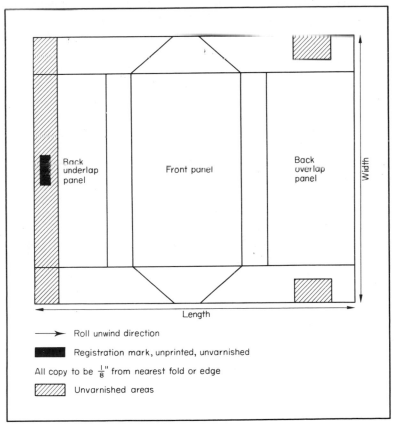

Figure 4.6   Layout drawing – toilet soap wrapper

applied to the pack. It also specifies any printing required for the packaging machine operations, for example, registration marks. Figure 4.6 shows a typical layout drawing for a toilet soap wrapper which is to be packed on a roll fed wrapping machine. The positions of the folds on the finished pack are marked and the distances of the copy from each fold line specified. Unvarnished areas where glue will be applied by the packaging machine are clearly marked and the position of the 'electric eye' registration mark is specified. The drawing also gives an indication of the roll unwind direction and if required the direction of print on the roll.

Copy The marketing department usually have the responsibility for co-ordinating the preparation of pack copy. The pack copy can be considered in three parts, advertising message, consumer information and legal requirements.

The advertising message, of which the basic brand name or logo are a major part, is devised by the marketing department usually in conjunction with their advertising agency. The amount of copy necessary is viewed in relation to the basic packaging design already chosen.

Consumer information takes up a large part of many packs. This may be in the form of recipes (food products), usage instructions, warning notices or assembly instructions. Consumer research evidence is studied when preparing this part of the copy to assess what type of information is required.

Finally, there is the legal part of the copy to be considered. This may cover content marking (weights and measures), formula information (labelling of food regulations) or poison notices (hazardous goods). Chapter Eleven covers these aspects in more detail.

Copy development is completed by circulation of a mock-up of the proposed pack to all the departments concerned for approval. A typical approval list would be:

1	Marketing	— Overall pack approval
2	Legal	— Labelling requirements
3	Research	— Consumer information and formula details
4	Packaging	— Pack layout and dimensions
5	Manufacturing/ Engineering	— Machinery requirements
6	Distribution	— Shipping/outercase markings

A form is supplied with the 'mock-up' for signature approval and notes of necessary modifications to be made.

Art-work Art-work is prepared for all packaging items following approval of the pack mock-ups. Preparation is usually carried out by the packaging material supplier's art department. The only exception to this being where detailed photograph or half tone work is involved, when the pack designer may prefer to prepare the art-work together with transparencies of the necessary illustrations.

The art-work is normally prepared to scale on a hard-board background with transparent overlays showing the colour separations. Approval in the user company is usually given by the purchasing and packaging departments. The purchasing department also has the responsibility of officially handing over approved art-work to the supplier together with a purchase order for the initial supplies.

Proof approval The finished art-work is used by the supplier to prepare plates, cylinders or silk screens for the finished pack decoration. The time taken for this process varies depending on the type of pack decoration. Cylinders for gravure printing can take up to two months, whereas silk screen or flexographic plates can often be produced within a few days. Proof samples from the prepared material are submitted by the supplier to the user for approval. Often these proofs are prepared on laboratory scale equipment and can be judged only for copy and layout. In these instances the usual procedure is for the purchasing agent to attend the initial production run to give final colour approval for the finished package. The printing of cartons by either gravure or flexographic methods is a typical example of this. Initial orders are produced by the supplier following proof approval on the machine by the purchasing agent.

Quality control procedures

Two types of quality control procedures are implemented prior to the start-up of a new pack, the first one governing the quality of incoming packaging materials, the second one controlling the finished product pack quality.

Packaging material inspection   A routine method of inspection of incoming packaging materials is agreed. The degree of inspection will depend on the overall quality policy adopted (see Chapter Seven for methods of controlling package quality). For the initial order, however, all the tests specified will usually be carried out. Arrangements are made for the supplier to send in a randomly chosen test sample prior to the main delivery. These pre-delivery samples are carefully checked against the specifications by both the quality control and packaging departments. This procedure gives an early warning of the potential problems with off-standard packaging materials. A method for the reporting of defective packaging materials via the purchasing department to the packaging supplier is also arranged.

Finished pack standards   Acceptable standards of finished pack quality are agreed with the manufacturing and quality control departments. These include both quantitative test methods such as the measurement of closure torque on glass containers and also subjective assessments of such features as overall pack appearance. In the latter case the quality control department will often draw up a rating scale to supply a quantitative assessment.

Start-up

Observance of the procedures outlined in the last two chapters should ensure that the start-ups of new packs or packaging developments are successful. Problems can still arise for a variety of reasons. Close attention is paid to all aspects of the packaging operations during the start-up.

Packaging materials

Problems with packaging materials are investigated and if necessary followed up with the supplier. One of the most difficult areas with start-ups is to track down the source of problems in the packing operation. Often packaging materials are blamed for poor machinery performance and vice versa. In some cases the initial start-up experience will indicate the need for changes in packaging material specifications, for example, slight changes in pack dimensions. Quality control test methods are also checked to confirm that these comply with the tests stated in the specification.

Machinery performance

The speed and efficiency of the packaging equipment is compared with the predicted rates. Minor problems are listed for future attention by the supplier's service engineer. Initially during the start-up period the overall machine efficiency will be affected by the lack of expertise of the operators and maintenance personnel. Depending on the complexity of the new equipment a minimum period of time is allowed before maximum efficiency is obtained. Contact is made with the supplier immediately for serious machine problems.

Line efficiency

After the packaging machinery has reached the expected efficiency, a line study is carried out to check on the labour content of the operation and compare it to the estimate on the standard cost sheet. Potential methods of improving overall efficiency are noted.

Finished pack appearance

The appearance of packs coming off the production line is assessed during the start-up period and the results compared with the official quality control assessment. Often as a result of experience during the start-up, quality control tests have to be modified to make them more realistic. This applies particularly to the tests measuring defects in pack appearance; rating scales often have to be modified depending on the type of defects experienced.

Market checks on the in-store appearance of the new pack are also carried out to confirm that the quality control methods are adequate. These market checks also serve as a final check on the suitability of the finished pack for the required distribution channels.

Effective communications play a major part in the success of any packaging development programme. In this chapter we have seen the role of the various departments involved with the execution of the development. Close contact between all these areas is necessary for effective progress. Equally important is close attention to detail at all stages of the work. Minor oversights, even at late stages in the development, can often ruin all the valuable work carried out earlier in the execution process. Prompt action is also necessary when problems arise to progress the development and to maintain the predicted time schedule.

5

Packaging Evaluation

Package evaluation falls neatly into two parts; technical testing and consumer evaluation. The latter has already been dealt with in Chapter Four; ('Approach to Packaging Development — Execution') and this chapter will deal with technical aspects of package testing.

Technical package evaluation has three main objectives:

1 To predict the performance of the package in use and to ensure that the pack is satisfactory under all conditions of use, at the minimum cost.
2 To control package quality.
3 To obtain comparative information, ie between different packs, or to obtain information about the specific strengths and weaknesses of a given package when exposed to an individual hazard.

How these objectives can be achieved is outlined in the following sections.

Test procedures

Whatever the reasons for testing, the testing itself can be divided into three main areas; simulated laboratory testing, field trials and production testing.

Before looking at any of them in detail it is worth considering the question of who should carry out the testing. Production testing, of course, must be carried out by the package user on his own filling equipment, but field trials and laboratory testing might well be done outside the company. One reason for keeping testing firmly inside the company is security; particularly in the case of new package developments. Other reasons include being able to carry out more accurate costing on a pilot scale before committing the project to production, being able to train production personnel on a pilot line and easier development of a new package specification before production. If security is not a problem it may be worth considering the use of a contract packer's facilities. In this way little or no cost is incurred for the installation of pilot equipment and no space is tied up within the company. In addition, most contract packers have well-equipped laboratories and quality control facilities.

Package evaluation facilities also exist within packaging material suppliers. As they are designed for the development of the supplier's materials (with particular reference to their suitability for the intended end-uses), they are likely to be helpful in the user's package evaluation programme. On the other hand, the supplier has to look to his own priorities and must concentrate on the large volume markets. On balance it would seem preferable for the small package user to use his own or a contract packer's facilities and remain independent of the packaging materials suppliers; but this may not always be possible.

Information collection

Before a meaningful test programme for any pack can be devised it is necessary to collect as much information as possible about the hazards that a pack will encounter during its life. Where possible, the information should be quantified but qualitative information on the type of hazards is also very useful. Some of the common hazards are discussed below.

Transport hazards

Transport hazards will, of course, depend on the type of distribution system used. In the case of rail transport, one of the main hazards is shunting shock. This occurs when trains are assembled in marshalling yards and sidings and the damage resulting from it will depend on the method of loading, the method of bracing the goods in the wagon, and the nature of the packed goods themselves. Shunting is not normally a serious hazard in the UK for light containers and goods of normal strength. If the packaged goods are particularly shock sensitive then the use of specially sprung vans may be advisable. Shunting can also cause trouble where sacks are concerned unless care is taken in loading. If shunting causes the load of sacks to slip and move against the sides of the van, the sacks will be torn when the sliding doors are opened.

Shunting shocks can be measured using mechanical recorders or by carefully positioned strain gauges.

Vibration can be another hazard, particularly to light engineering equipment and domestic appliances. This can cause loosening of screws and nuts or movement of parts which should remain immovable. The nature of any vibration will depend on the type of van used, the condition of the track and the speed of the train. It is particularly important to consider packaging at the product design stage as this can often obviate the need for expensive protective packaging.

Vibration is also a hazard with road transport. The type of vibration will depend on the condition of the road, the speed of travel and the loading and spring characteristics of the particular vehicles. Also, bouncing of the load may occur due to major irregularities in the road surface. Although this is not a matter of great importance in the UK, it must be considered when exporting to some of the less well-developed countries.

Simple mechanical recorders are usually unsatisfactory for measuring road and rail vehicle vibration since adequate sensitivity is difficult to obtain in the case of higher frequency vibrations.

Electromagnetic accelerometers are often used and can be used in conjunction with high-speed pen recorders. If the signals are amplified and then recorded on magnetic tape they can be subjected to a detailed analysis.

Transport by ship may present hazards due to water (when the goods are carried as deck cargo) or to pitching and rolling especially when goods are stowed in stacks of 20-30 feet (6-9m.) high. There may also be vibration due to the engine or propellers.

Air transport possesses few major hazards, the main ones being the low temperatures and pressures encountered in the holds. However, since light weight is extremely important in the case of air travel, packaging is normally reduced to a minimum and it is important to take account of any other transport which may be used at the end of the journey to take the load from the airport to the final destination.

Loading and unloading hazards

These handling hazards can occur whatever transport system is used. They may be encountered not only at the manufacturer's works and at the final destination (at a retail shop or factory), but also at transfer points *en route*. Such transfers include movement from one railway van to another at a station or siding, from one lorry to another, and from road or rail onto a ship at the docks.

The two main hazards are drops and impacts of one container against another. The latter occurring when containers are thrown or dropped by a crane onto other containers during stacking.

Drop height and drop frequency will depend on the handling facilities available. Ideal conditions are the provision of hand trucks or mechanical handling facilities and the existence of loading platforms at the same level as the lorry or rail floor.

Drop hazards also depend on the weight of the package being handled. Packages which can just be handled comfortably by one man will normally receive better treatment than those which are heavy and awkward to carry. Equally, packages which are too light are liable to be thrown around a great deal. The preferred range is around 30-50 pounds (13.6-22.7kg). Containers weighing around 300 pounds (136kg) are likely to be rolled or pushed out of the lorry or van, particularly if they are awkwardly shaped. Heavier packages, those around the 500 pounds (227kg) mark, will usually require some sort of mechanical handling equipment and the drop hazards should therefore be less. Once again it is necessary to distinguish between developed and under-developed areas of the world. It is still possible to find many regions where mechanical equipment is unobtainable and heavy packages must, perforce, be handled by four or five men at one time. Drop hazards are, therefore, much increased in such regions.

One very simple shock recorder is the PIRA Journey Shock Recorder which consists of a flap restrained by a spring. When the flap receives a blow in a given direction it operates a counter. By adjusting the spring tension, the instrument can be made to count once each time the pack is dropped on a given face, from above a given height. A number of these drop counters can be placed in a container.

More complicated instruments are available which can record drops on more than one face and can also indicate the timing of the various drops.

Climatic hazards

Climatic hazards will be determined not only by the final destination of the goods but also by their route and method of transport. The main ones are changes in temperature, exposure to liquid water, exposure to harmful humidities and exposure to light or direct solar radiation.

Changes in temperature   Exposure to high temperatures will accelerate corrosion, other chemical reactions and biological changes. In many cases, therefore, low temperatures can be beneficial as they slow down chemical and biological changes in the product. However, there are many products which are adversely affected by low temperatures, as mentioned in Chapter One (for example, emulsion paints and adhesives). Changes in temperature may also be harmful even though the extremes of temperature are not in themselves liable to cause deterioration. The cooling of warm, moist air leads to condensation of water vapour, often inside the package itself.

Exposure to liquid water   Liquid water in the form of rain is a very common climatic hazard. Once again the route must be taken into account as well as the final destination. In many cases rainfall is seasonal and it is essential to obtain rainfall figures from the areas concerned, for each month of the year.

Exposure to harmful humidities   The question as to whether a particular humidity is harmful or not will depend on the nature of the product. Low humidities cause drying out of moist products, the effect being one of humidity equilibrium between the inside and outside of the package. If the relative humidity of the inside of the package is the same as that of the atmosphere then no changes occur. Products such as tobacco, cakes, soap, etc deteriorate badly when moisture is lost and must, therefore, be well-protected if exposed to a dry climate or if they are likely to spend any appreciable time in one.

High humidities are, of course, harmful to dry products which cake, deliquesce or become sticky when damp.

Effect of light or direct solar radiation   Many products are adversely affected by light. The light in certain export countries may be more intense than in the UK and may also contain a greater amount of ultra-violet light. The effect on the product (if unprotected) may be changes in colour or a catalytic one on some chemical reactions such as oxidation of a fat, leading to rancidity. Light may also have an effect on the package; discolouration of package decoration may occur. Some plastics are also adversely affected by ultra-violet light. Polyethylene and polypropylene, for example, are liable to embrittlement by ultra-violet light.

Other hazards

These include micro-biological hazards, insects, rodents, etc. Many of the materials used in packaging are susceptible to attack by micro-organisms and by insects. For example, fungi and bacteria grow on animal glue and starch adhesives and can also grow on damp paper and board. Attack by fungi can also occur even on materials which are not themselves nutrients. Thus, aluminium is attacked by acids given off by certain fungi, while glass is attacked by enzymes produced by moulds. The micro-organisms are nourished in these instances by impurities on the glass and aluminium so that cleanliness is the answer to the problem. Because fungi need moisture, the hazard is greatest in tropical countries.

Insect infestation is another hazard particularly in the case of wood, paper or fibreboard packaging. In general, warm conditions favour insect growth, and storage in warm, damp conditions should be avoided.

Finally, rodents may be a hazard in the holds of ships and warehouses. Nearly all packaging materials are susceptible to rodent attack, even tins having been eaten through by rats.

The data needed before tests are devised for any particular package are given in Figure 5.1

Simulated laboratory tests – unit containers

Unit containers cover a wide range and tests vary considerably according

Mechanical hazards	
Hazard	Date needed
Drops	Position of impact on package. Height of drop. Impacting surface
Impact with other packs, etc	Type of impacting surface (ie vehicle, dock, warehouse, other package, etc)
Crushing	Stack height. Nature of nets (if applicable). Material of slings (if applicable)
Vibration	Amplitude, frequency, whether continuous or intermittent, whether packages under a top load
Climatic hazards	
Hazard	Data needed
Liquid water (rain, spray, etc)	Amount, and time of rainy season
Humidity	Relative humidity measurements of warehouses, transport, etc
Temperatures	Magnitude and variations likely
Solar radiation	Intensity and type (eg amount of ultra-violet light present)

Figure 5.1 Distribution hazards

to the type of container. The tests are, therefore, discussed under the various container types.

Glass containers

With the growth of high speed filling lines the dimensions of glass bottles have become of increasing importance. On modern filling lines too great a variation in height, body diameter, (or equivalent dimensions of non-cylindrical bottles) and neck finish between containers, can lead to damage by star wheels, filling heads or capping machinery. The checking of body dimensions is usually carried out using 'go' and 'no go' gauges. The overall diameter of the neck finish can also be determined in this way but checking of the contour is carried out by using a shadowgraph which projects an enlarged profile of the finish against the specified contour.

Modern weights and measures legislation has also made it important to know the accurate capacity of a container. Capacities may be measured 'brimfull' or to a stated height. Because of manufacturing variations a number are selected at random and the mean capacity measured.

Tests designed to measure the resistance of the bottle or jar to mechanical hazards include:

1 Drop tests.
2 Pressure tests; especially for containers to hold carbonated beverages (or aerosols).
3 Impact tests; to test resistance to impact on filling lines or during handling.
4 Tests for thermal shock. These are important when glass containers have to be filled with hot products. The bottles are immersed in hot water for a given time then plunged into cold water. The test is carried out from hot to cold because this causes a greater effect than from cold to hot, hence a safety factor is automatically incorporated.

Metal containers

Tinplate and most steel containers are made in more than one piece. If the containers are to hold liquids or semi-liquid products the containers are tested for leakage under air pressure, before filling. Aluminium containers are normally made in one piece by an impact extrusion process.

Certain containers, such as some tobacco tins, are vacuum sealed and should be held for a period of time after filling to check for vacuum retention.

One of the most important items to be checked with metal containers is liability to corrosion. Internal corrosion (by the product) is assessed by storing filled containers under room temperature and elevated temperature conditions and opening containers at intervals for visual examination. Results of elevated temperatures can be used as accelerated tests to predict results at room temperature conditions, as well as for actual results under tropical conditions.

External corrosion, by ambient conditions, is assessed by storing filled cans under conditions approximating as closely as possible to the conditions likely to be encountered by the commercial packs. Filled tins should be used because the heat content of the pack determines its speed of response to changes in temperature and thus its liability to condensation.

A more specialised test, applied to the wax coated lead tubes used for packaging fluoride-containing toothpastes, is designed to detect pin-holing or other damage in the wax coating. The tube is filled with an electrolyte and electrodes arranged so that a current can only flow if the wax coating is discontinuous.

Films/laminates

Some of the major tests for films and laminates are outlined below.

Seal strength   One of the most important properties on wrapping equipment or form/fill/seal machines is the seal strength. This is usually determined by measuring the force required to pull apart the pieces of film which have been sealed together. A strip is cut through the heat seal and the two free ends are placed in the clamps of a tensile testing machine. The force necessary to peel apart the two pieces of film is then measured. A static test is also sometimes used. The strips are hung from a frame with one end clamped and the other attached to a weight. The seals are examined at intervals for signs of failure. When recording the results of this test, the weight and the length of time the load was in operation are both noted.

The seal integrity of complete sachets or pouches can be measured by immersing them in water contained in a vacuum desiccator. Any leakage is immediately apparent from the air bubbles forming at the site of the leakage.

Water vapour transmission rate (WVTR)   This is a measure of the permeability of the film to moisture vapour under controlled conditions of temperature and humidity. There are two main methods: weight gain and weight loss methods.

The weight gain method utilises shallow dishes, carrying a desiccant and covered with the film under test. The edges of the film and dish are sealed with wax. The dishes are weighed initially and then placed in a temperature and humidity controlled cabinet. Weighings are carried out at intervals and the weight gain measured.

In the weight loss method a metal cup, containing a standard quantity of water, is sealed by the film under test. The sealing medium round the edges of the film and cup is again wax. The cup and contents are weighed initially, then placed in an oven maintained at a given temperature and humidity. The cups are reweighed at intervals, the loss of weight being a measure of the transmission rate.

Instruments are also available for the automatic measurement of WVTR.

Gas transmission rate   The measurement of gas transmission rates is carried out under controlled conditions of pressure, as well as of temperature. Essentially the normal test consists in making the film a partition between a test cell and an evacuated manometer. The pressure across the film is usually controlled at one atmosphere. As the gas passes through the film sample, the mercury in the capillary leg of the manometer is depressed. After a constant transmission rate has been achieved, a plot of mercury height against time gives a straight line. The slope of this line is used to calculate the gas transmission rate.

Closures

A closure must provide a positive seal and yet permit the easy opening of the container. In many cases it must also re-seal the container after use. There are so many different types of closure in use with many different performance requirements that it is impossible to cover the testing of closures in part of one chapter. Described briefly below is one of the most common types of closures; the pre-formed screw threaded cap.

Shadowgraph   The shadowgraph is used in an analogous way to that mentioned for examining the contour of a glass bottle finish. The closure is cut in half and placed on the shadowgraph which projects an enlarged image of the contour of the thread finish. This method of measurement is, of course, useful for examining the contour of other types of cap such as snap-on ones, or plug fittings.

Torque retention   Torque is the applied twisting force necessary to tighten a closure on a bottle or remove the tightened closure. In general, opening torque is lower than the closing torque. The retained torque must, of course, be high enough to prevent the closure coming off during transit and storage. Torque retention is measured by clamping the closed bottle in a spring loaded mechanism and twisting-off the cap. The force necessary is measured on a dial by a pointer attached to the clamp.

Breaking torque   This is the force required to cause the closure to crack during application. It is important since it is a measure of the closure's ability to stand up to the closing torques on automatic capping equipment.

Paper/board

Tests for paper and board can be divided into strength measurement, barrier properties and surface appearance (including printing). Some of the more important are given below.

Burst strength (Mullen test)   The paper or board is clamped into position over a circular rubber diaphragm. Air or hydraulic pressure is applied to the underside of the rubber diaphragm until the sample ruptures. The pressure at that point is indicated on a gauge. The results indicate the resistance of the board to puncturing.

Tear strength   One edge of the sample is held in a stationary clamp, the opposite edge being held in a moveable clamp attached to a pendulum held in a raised position. When the pendulum is released, some of the energy is lost in tearing the sample and the amount is registered by a pointer on a quadrant scale. The sample is nicked before the test is commenced in order to initiate the tear.

Compression   This test is carried out on a sample strip bent round into a circle and placed on edge on the bottom plate of a hydraulic press. The top plate is brought down onto the ring sample until it is collapsed. The maximum load is measured and recorded as the ring crush strength. The test is a good indication of the stackability of the filled container.

Stiffness There are several instruments available for measurement of the stiffness of paper and board. Taber and Gurley are the names of two typical ones. These consist of applying a standard weight to a sample strip of board and measuring the deflection when the weight is removed. It has proved difficult to correlate results from different instruments and also to tie in test results with performance in use or on machines.

Water vapour and gas transmission rates These are measured in a similar manner to that described above for films and laminates.

Grease resistance The sample is placed on a piece of book paper and a weighed amount of sand is placed on the sample, using a glass tube to ensure that the pile of sand has a uniform area. Coloured turpentine is added to the sand pile using a pipette and the time required for a stain to appear on the book paper is measured.

Resistance of print to rubbing This can be measured on automatic instruments or by simple manual tests. A sample of the printed material is brought into contact with either a piece of corrugated board, or a second piece of the test material and a given weight is applied. A number of standard rubs are then carried out and the surface appearance examined after rubbing.

Resistance of print to product Ink resistance tests are carried out by applying a given amount of the product to a sample of the package so that the product is in direct contact with the inks. The samples are then stored at elevated temperatures to check for any reaction between the two.

Resistance of print to light Accelerated tests for light resistance rely on the use of light from a carbon arc or a Xenon lamp. The test is useful for comparison purposes between different printing inks but it is not easy to correlate the results with those obtained using direct sunlight. Modern instruments, such as the Xenotester, attempt to simulate typical day/night conditions and variations in climatic conditions.

Gloss Gloss is measured on a glossmeter which consists basically of a light source and photosensitive receptor. The light shines on a sample of paper or board at a specified angle and some of it is reflected onto the receptor. The fraction of the original light which is reflected is the gloss of the sample.

Brightness   A similar instrument to the glossmeter is used and the reflectance of a piece of paper or board is compared to that of a standard white block (usually magnesium carbonate). Test results are recorded as a percentage of the standard material.

Plastics containers

The main tests carried out on plastics bottles are, drop tests (for measurement of impact strength), tests for the effect of the product on the bottle and measurement of permeability.

Drop tests   There are many different test procedures. In one, the bottles to be tested are filled with water and dropped down an un-plasticised PVC tube onto a level, rigid baseplate. Two tubes are usually used. One is 8 feet long and is drilled to allow drop height variations of 6 inches, while the other is 4 feet long and is drilled with holes 3 inches apart. A peg is placed through a hole of the required height and the bottle placed in the tube. Removal of the peg allows the bottle to fall freely down the tube. The results of dropping a large number of bottles from a large height can be treated statistically, or the drop height can be altered according to the result of the previous drop.

Effect of product on plastics   Basically, tests are carried out by half filling the bottles with product and storing at ambient and elevated temperatures. Damage may be assessed visually or by carrying out drop tests on the bottles after storage. One important effect which can occur with certain plastics such as the polyethylenes, in contact (under stress) with polar organic compounds such as detergents, is environmental stress cracking (ESC). Typical test methods are described in Chapter Four.

Permeability   The bottle under test is almost filled with the product and securely sealed with a stainless steel insert and an oil-resistant rubber O-ring which is placed in the bottle neck. A polypropylene cap is then screwed on at a constant torque. The total weight of the assembly and

the contents is recorded and the weight of the contents alone is determined. Six bottles are usually sufficient; three being stored at 23°C and three at 50°C. Percentage loss in weight is calculated for each bottle, together with the mean percentage at each storage condition (ignoring bottles with excessive weight losses due to faulty sealing). The permeability factor P is calculated as, $P = \dfrac{WL \times T \times 100}{A \times D}$
where *WL* = weight loss in grams.

T = Mean wall thickness of bottle in micrometres

A = Internal surface area of bottle in square centimetres

D = Storage time in days

Simulated laboratory tests — outer containers

Tests for outer containers may be designed as comparative tests, in which case the containers are usually tested to destruction, or to simulate actual conditions. Some of the more important are described below.

Drum test

The drum is six-sided and is rotated in a vertical plane about a horizontal axis. The internal faces of the drum are fitted with baffles and guides which ensure that the package inside is turned to give various positions of impact. Drum tests give reasonable results when comparing one package with another but correlation with actual conditions is poor.

Vibration tests

These are designed to simulate the effects of the vibration experienced in road or rail travel. The apparatus used is a vibrating table which consists of a bed, driven by two eccentrics, one at each end, connected in phase with one another. A platform is attached to the top of the vibrating bed and this undergoes a circular harmonic type of vibration when the equipment is in use. The amplitude is usually fixed at 1 inch (25mm), while the frequency can be varied to cover the range of frequencies encountered in railway vans or road vehicles (120 cycles a minute to 360 cycles a minute). A rough correlation between the vibrating table and use has been given as one hour on the vibrating table is equal to a 1,000 mile journey by rail.

Inclined plane test

This is a means of reproducing shunting shocks and consists of a track inclined at 10° to the horizontal on which a wheeled platform can be released so as to impact wooden buffers at the other end of the track. The speed at which the truck is moving at the moment of impact can be varied by altering the starting positions along the inclined plane. Speeds up to about 8 miles (13km) an hour are normally used. The effects of the shocks experienced by the package are related to shunting shocks or to side impacts during transit.

Drop tests

The apparatus needed for drop tests is simple but effective. There are two main types in use. One consists of some form of release mechanism whereby the package is suspended by means of a sling which allows it to be dropped from a selected height, in any particular position, on to any type of floor.

The other type of equipment is the table drop test. The package is held in the desired position on the table top and a trap door is opened, thus allowing the package to fall to the floor. The height, and position, of fall and the type of floor can again be altered as required.

Stacking test – static load

In this test the package is placed under a static load equivalent to the weight it would experience from all other packages directly above it in a stack of height equivalent to that used in practice. The package is examined at intervals for assessment of damage, to the outer package and to the contents.

Compression test

The stacking test is likely to be a time consuming process, unless the packages are totally unfit for their job, and reasonable estimates of the safe stacking load for any crushable container can be obtained more quickly from a dynamic compression test. The package is placed between the two platens of a hydraulic press and a steadily increasing load is applied. The safe stacking load is usually taken to be between one quarter and one third of the load which the package fails in the compression test.

Test sequences

In any distribution system the main mechanical hazards can be divided into drops and impacts, compression in stacking, and vibration. From the data obtained concerning the number and magnitude of these hazards in a particular distribution system, it is possible to devise a sequence of tests which will determine the suitability of any particular package for that distribution system.

Interpretation of the results of a test schedule will depend a great deal on the experience and knowledge of the testing organisation, and suggested schedules should be taken as a guide only, and not as a rigorous specification.

Two examples of typical test schedules are given in Figures 5.2 and 5.3

Field trials

These should provide a link between the laboratory tests described earlier and the performance of the package in practice. There are many factors to be considered when designing a field trial.

1 The hazards on any particular journey will vary.
2 They will also vary when the starting point and destination are changed, even if the type of transport used remains the same.
3 Personnel handling packages become used to handling a new pack and so the results on a first trial may be different from those obtained on subsequent trials.
4 Any special treatment or arrangements made for observation of a trial may affect the handling and make it different from that which will be used when no observers are present.

TEST SCHEDULE B

(Home trade including Northern Ireland)

Weight 9 - 45kg maximum dimensions 910mm x 610mm x 610mm

Principal methods of distribution:
1 Passenger train
2 Mixed goods — road/rail
3 Full container loads — road/rail
4 Manufacturers' own transport

5 complete packages required for the tests.

Package No. 1: 300mm drop test onto
    (a) Base
    (b) Longer base edge
    (c) Shorter base edge   ) Diagonal
    (d) Opposite base corner } Horizontal
    (e) Top   )

Vibrating table test: Under load equivalent to a stack 2.40mm high,
                    15 minutes on base.

Stacking test: Under load equivalent to a stack 3.60m high
                    (not less than 115kg) for 24 hours.

EXAMINE

Package No. 3: As for package No. 1, but top used as base
Package No. 4: As for package No. 2, but larger side face used as base
Package No. 5: Drop test as for package No. 1
              Inclined Plane test: 300m run — one blow onto each
                                side face and vertical edge
              Drop test: 600mm drop onto
                            (a) Base
                            (b) - (e)   Each base edge in turn
                                         (diagonal horizontal)
                            (f) Top

EXAMINE

ASSESSMENT:
1 If all five packages pass satisfactorily, they should be suitable for all
   methods, including passenger trains.
2 If package 5 fails, it is not suitable for general distribution by passenger
   train.
3 If packages 3 and 4 only fail, then package is suitable for all forms
   provided it travels on its base.
4 If only pack 1 is satisfactory, then it is suitable only for good handling
   distribution system.

Figure 5.2 Test schedule B

TEST SCHEDULE D
(Export — road/rail and sea)

Weight 9 - 45kg Maximum dimensions 910mm x 610mm x 610mm

5 complete packages required for the test.

Package No.1:   450mm drop test onto
        (a) Base
        (b)-(e) All base edges in turn ⎫
        (f) Base corner   ⎬ Diagonal
        (g) Opposite base corner ⎭ horizontal
        (h) Top

Inclined Plane Test: 3m run. One blow onto each side face, and
                      each vertical edge (8 blows)
Vibrating table test: Under load equivalent to a stack 2.40m high,
                      30 minutes on base
Stacking test: Under load equivalent to a stack 5.40m high
                      (not less than 224kg) for 48 hours.

EXAMINE

Package No.2:   750mm drop test onto (a) Base
                              (b)-(e) All base edges in turn
                              (f)-(i) All top edges in turn
                              (j) Top
              300mm drop test onto (a)-(d) Each side face in turn

EXAMINE

Package No.3:   As for package No.1, but top used as base
Package No.4:   As for package No.1, but larger side face used as base
Package No.5:   Shower test: Hold under shower for 10 minutes,
              followed by storage at 38°C and 90% R.H. for 7 days

              Drop test: From 450mm as for package No.1

              Stacking test: As for package No.1

EXAMINE

ASSESSMENT: 1 If all five packages pass satisfactorily, they should
                 be suitable for all systems.
           2 If all packages pass except No.2, they should be
                 satisfactory where handling is good.
           3 Failure in packs No.3 or No.4 — only suitable where
                 pack is likely to travel the right way up.
           4 Failure in pack No.5 only — suitable where no
                 shower or humidity hazard.
           5 If all packs except No.1 fail — only suitable where
                 good handling can be ensured.

Figure 5.3 Test schedule D

It is necessary, therefore, when designing a field trial to use a number of different routes and send a number of consignments over those routes at various intervals of time. A pack of known performance should also be used as a control and observations should be made of the handling at various stages of the journey.

Machine trials

Machine trials are of equal importance to laboratory tests and field performance trials. This is obvious in the case of a container which is impossible to fill on the existing line and so incurs major capital expenditure, but is equally valid for somewhat smaller defects which will either slow down the filling line or increase the down time through stoppages.

The number of factors controlling performance of the package on the filling line (and, of course, handling before and after filling) depend on the type of package material, the product to be filled (chemical constitution, physical type, eg, liquid, powder, granule, etc) and the type of equipment (wrapping, form/fill/seal, case packer, etc).

Some of the performance factors to be considered are listed below.

Speed of filling

In the case of tins and bottles the speed of filling may be affected by the neck diameter. If this places a size limit on the filler orifice, it may be possible to attain satisfactory filling speeds either by increasing the flow through the orifice or by having a multi-head filler. Liquids liable to foaming will have this tendency increased by increasing the velocity of flow through the filler.

Heat sealing

Efficient heat sealing is dependent on the type of material, the pressure, temperature and dwell time; the last three being machine variables. A change in packaging material may necessitate a completely different type of heat sealer but even minor changes in package type should be checked for their effect on heat seal efficiency.

Accuracy of filling

A change from a rigid to a flexible bottle may make it difficult to use vacuum filling equipment but even if collapse of the bottles does not occur, the accuracy of fill must be checked.

Appearance

The appearance of cartons in particular may be damaged by passage through the filling machine. This can be caused by scuffing (by conveyor rails), or by carton erecting mechanisms. Previously satisfactory performance by other packages does not necessarily mean equivalent performance by the new one. Slight increases in dimensions could lead to abrasion of decoration by guide rails or the decoration could, of course, be less resistant to the scuffing action. The latter should have been shown up in laboratory tests.

Seal efficiency

This has been touched on in the case of form/fill/seal or wrapping equipment where heat sealing is involved. Screw capping must also be checked since a change in bottle or closure will need a change in machine conditions.

Other examples of the effect of packaging changes on filling equipment were given in Chapter One, 'Introduction to Packaging'.

The main target for any type of testing, whether it be laboratory assessment, field trials or production trials, must be the correlation of such testing with actual behaviour of the package in use. This means that tests must always be carried out on production samples. Prototypes are useful in the design stage and may provide information which is useful in modifying the design but no final decision can be made until machine-made samples have been tested. Similarly, the use of simulated laboratory tests can provide a great deal of useful information but this must be confirmed by field trials as far as possible.

This chapter gives guidelines on how to set up test programmes. For practical details of the various tests mentioned (and others), the test specifications of the British Standards Institution (BSI) or the American Society for Testing Materials (ASTM) should be consulted.

6

Packaging Specifications

A packaging specification is a written statement of the requirements for a given packaging material item by the user to the supplier. It communicates the technical details of these requirements to the supplier and also sets out the criteria for determining whether the requirements are met. It is important that it is clear and concise and that it is written in terms common to both the supplier and the user. A comprehensive packaging specification system forms the basis of any efficient packaging/purchasing operation.

Purpose

Packaging specifications are normally produced by the department responsible for the technical aspects of packaging. This may be an individual packaging department or a unit of research and development, quality control or manufacturing departments (see Chapter Two 'Organisation of the Packaging Function'). The main objective of the packaging specification is to ensure that the nature of the packaging of the finished product satisfies the original marketing brief. This applies both to the finished appearance and to technical criteria. The packaging specification does, however, have many secondary functions to both departments within the user firm and to the package supplier.

Packaging supplier

The supplier checks his own process specifications against that of the user's specification to ensure that he can produce material in compliance with the specification. The degree of accuracy demanded by a given specification will affect both the type of equipment on which the item will be produced and also the frequency of quality control checks. For example, the requirements for a paperboard carton will be much stricter for one which is to be machine packed at 200 per minute than for one which is hand formed and sealed. To ensure compliance with the specification the supplier can also use the information contained therein to set up or adjust quality control tests. Finally, it is used as a basis for the quotation to the user.

Purchasing department

The purchasing department orders packaging material items against a given specification. Often copies of individual specifications are attached to the purchase orders, particularly for new packaging items. Reference to a specific packaging item by its specification number and title on a purchase order can simplify issuing of these documents. It is important to ensure, however, that specifications are kept up to date and also issued promptly to the interested suppliers. The purchasing department has the responsibility for issuing specifications to the suppliers.

In companies where packaging specifications are not kept up to date, or indeed do not exist at all, the purchase order is often the only place where the packaging material description occurs. Usually it is left to the regular supplier to continue supplying items as for previous orders. This can lead to a gradual drift downwards in quality. For example where colour standards are not enforced and not kept up to date, significant changes in a colour shade can occur. Difficulties also arise when a change in packaging supplier is made. Information critical to the new supplier may not appear in what are often perfunctory details on the purchase order.

The purchasing department also use the packaging specifications to give to new or alternative suppliers as a basis for their quotation. A readily available specification sheet saves time in terms of verbal and written explanations.

Engineering department

Performance of the packaging material item on packing and filling equipment is one of the main requirements included in any specification. The engineering department can adjudge from a packaging specification the expected machine performance and will use the packaging material specification when purchasing new equipment. For example, an engineer purchasing a new form/fill/seal machine would supply both a sample of the packaging material and a detailed packaging specification to the machinery supplier. Reference to the specification is also made on the purchase order for the machine.

Quality control department

Quality control checks carried out on incoming packaging materials are based on the information and test methods contained in the packaging specification. The quality control department is often called in to investigate packing problems and to decide whether the packaging material or the machinery is at fault. This is an extremely difficult area where experience plus careful analysis pays off. The natural reaction during machine breakdowns and problems is often to blame the packaging material. In many cases faulty machine setting turns out to be the cause of the problem.

Investigation of packing problems in the user's plant often leads to suggestions by quality control or manufacturing personnel for changes in packaging material specifications which will improve overall productivity. These may concern slight alterations in, say, dimensions or more major changes such as in material type.

Manufacturing department

Certain items from the packaging specification are taken and included in the manufacturing packing standards. These lay out the details of the units produced on a given packaging line. For example, from a roll stock specification for flexible packaging, the manufacturing department would extract details of sealing temperatures, pressures and expected packaging material loss. The specification would also give information on any control checks which may be part of the manufacturing operation eg 'leaker' tests on finished pouches.

Design department

Layout drawings from the specification form the basis for packaging artwork prepared by the design department. The packaging specification also provides details of print areas and accuracy of registration between colours. Standard colour charts, used as part of the packaging specification, are approved by the design department as complying with the required target.

Distribution department

Details of outercontainer dimensions and gross and net weights are used by the distribution department for the calculation of freight charges. These are obtained from the packaging specifications. The packaging department and the distribution department work closely to develop pallet patterns for packed stock. Details of pallet patterns often accompany the packaging specifications.

Packaging department

This is the focal point for the packaging specification system. The packaging department uses this system as a formal method of communication to all the areas involved with packaging to keep them advised of the latest packaging material requirements. A master set of all current and superseded specifications is maintained in the department for reference purposes. The packaging department sets up and maintains the system and ensures that new and changed specifications are issued promptly. When a large number of specifications are handled regularly a set routine is established for their preparation and issuance, for example part of one day at the end of each week is set aside for dealing with the week's issues.

Types of specification

In the packaging industry today there is no generally accepted method for the format or type of contents necessary, for a packaging specification. The layouts, methods of presentation, amount of detail and lengths of specifications vary widely from company to company. There are however two basic types of packaging specification namely performance and material specifications. All user company specifications can be described either under one of these headings or more often as a combination of the two.

Performance specification

A performance specification lists the packaging material requirements in terms of the desired end result rather than the type of material to be used. The desired end result may be expressed in terms of a standard test, a specially developed performance test, or on the basis of machine performance. Some typical examples of this type of specification follow:

1 Corrugated cases are specified in terms of compression strength rather than board composition.
2 Laminates are specified in terms of Moisture Vapour Transmission Rate (MVTR) instead of material components.
3 Carton-board is specified in terms of 'bulge' measurement instead of caliper or weight.
4 Plastic containers are specified in terms of drop test performance instead of weight or minimum wall thickness.

This method allows the packaging material supplier freedom to decide the type of materials used and to experiment with more economical alternatives within the performance limits. It allows for competition between suppliers to produce the required quality at the lowest price. It can also be of assistance in reducing delivery lead times. Implementation of a performance specification calls for close co-operation between supplier and user. In some cases supplier and user work jointly with the supplier's supplier to implement a performance specification.

An example of this was the development of a performance test for measuring the resistance of soap and detergent cartons to bulge. This particular performance factor was considered to be the most important feature of this type of carton as excessive bulge leaves a large headspace at the top of the carton, as well as causing filling machine problems. Co-operation between user, carton suppliers, board suppliers and equipment manufacturers resulted in the development of a 'bulge' measurement on specially developed Bulgeometer equipment. This equipment measures the deflection of a piece of carton-board when a given air pressure is applied. The test conditions being devised so that they are equivalent to those experienced by a soap or detergent carton either during machine filling or when fully packed.

Material specifications

Material specifications describe in detail the types of material needed to meet the required performance. By this method the user ensures that, providing processing and printing are satisfactory, the correct material performance will be achieved. To specify in this manner the user needs to be familiar in detail with the raw materials used by the packaging supplier. The material specification is based on experience using the specified materials or on simulated tests which are typical of the complete packaging cycle.

A material specification for a glass container for a liquid food product will list:

1 Type of glass
2 Colour of glass
3 Full dimensions and tolerances
4 Container weight
5' Container capacity
6 Type of container coating

When operating with material specifications of this type, it is important for the supplier to keep the user advised of any proposed raw material or processing changes and to receive authorisation prior to introducing the change.

Performance or material specifications?

Which type of specification is recommended for an ideal operation?

In practice most good specifications are a combination of performance and material requirements.

At first sight, the pure performance specification seems to be the ideal solution. It is flexible, it encourages the supplier to use his knowledge and experience to reduce costs and it delivers the required performance. The main problem occurs in trying to specify performance aspects exactly enough to ensure that the finished specification covers all contingencies. It is possible that the suppliers will develop materials meeting a set performance specification but also having some undesirable negative characteristics. An example of this is the procurement of an outercase for packaging of cartons of soap flakes. The susceptibility to mould growth of the board used for the outercases in the presence of high humidity (caused by the product) makes it necessary to limit the adhesive used to one which is resistant to mould growth. This is in addition to the normal performance requirements of outercase compression resistance.

In drawing up a material specification the user takes the required performance and applies his experience of materials and testing to list the requirements. The material specification tells the supplier exactly which components to use and does, therefore, tend to rule out innovation and cost savings by the supplier. This can be overcome, however, by the user continually appraising his specifications and discussing ways of improvement with his own departments and the supplier.

The type of specification to be used also depends on the relative volume and cost importance of the packaging item being considered. Performance specifications, which often require elaborate test method development, are often more suitable for packaging material items that are ordered in large quantities.

A specification that is typical of one including both material and performance criteria is that for a polyethylene liquid detergent container. In this case the type of material to be used is specified, (ie high density polyethylene) but details of the raw material supplier and grade to be used are left to the supplier. Performance tests for impact strength and resistance of the container to stress cracking are also included.

Specification system

One of the main problems encountered in operating a packaging material specification system is the time taken up in preparing and updating it. This problem can be reduced considerably by splitting the specification system into two groups: general specifications and detail specifications.

The general specifications cover the main bulk of the packaging material requirements and change only occasionally. The shorter detail specifications cover specific descriptions of a given item and are relatively simple to change.

General specifications

These are prepared for each type of packaging item. They cover the general specifications and allowable tolerances for that item. A typical group list of general specifications for a company producing consumer goods is as follows:

G-01	Bags, balers, and envelopes
G-02	Canisters
G-03	Metal cans
G-04	Closures
G-05	Cartons
G-06	Drums
G-07	Glass containers
G-08	Labels
G-09	Corrugated outercases
G-10	Roll stock (flexible)
G-11	Plastic containers
G-12	Aerosols

A common specification layout is used for all the general specifications (see Figure 6.1). This figure shows all the headings on a single sheet. Normally the length of the general specification will run to several pages. The requirements are then outlined as follows.

Scope   This simply states that the specification covers all the required construction and performance requirements for the package type described, and lists the range of packs covered.

Construction   This section covers all the requirements regarding mater-

Company Name Packaging Material Specification		
No: G	Type: General	Date:
Supercedes :	Material :	Distribution :

1 Scope

2 Construction

3 Performance

4 Appearance

5 Shipping

Reason for :   Approvals _____
  Change

                                        _____

                                        _____

                                        _____

Figure 6.1 General specification

ial and dimensions which are common to all items of the group.

In a general specification for flexible packaging roll stock, for example, tolerances for basis weight, slitting, colour register and methods of roll joining are included. Accepted tolerances, as specified by the relevant British Standard or by the industry association, should be used wherever possible. This may not always be possible, however, a good example being packaging materials for use on high speed packaging lines. In such a case, where it is necessary to specify closer than normal tolerances, agreement with the supplier must be negotiated prior to finalising the specification. Care should always be taken to ensure that demands are not too strict for the required performance, as overspecification inevitably leads to paying too high a price for a given packaging item.

Any other particular construction requirements should be included in this section. A food company, for example, might include a statement that all materials used are to be free from substances which may impair the flavour, odour or bacteriological status of the food to be packaged. A section should also be included stating that no change in construction material may be made without prior company approval.

Performance This section covers the specific performance requirements applied to a group of items. It gives details of methods of testing and also of allowable tolerances. For specifications which are worded more in performance terms this is obviously a vital section. Performance tests listed in this section should include the following information:

1 Preconditioning
2 Test conditions – temperature, humidity etc
3 Test equipment
4 Number of samples to be tested
5 Frequency of testing
6 Methods of testing
7 Recording of results
8 Allowable tolerance on results

Detailed investigation of packaging material complaints can often lead to suggested improvements in performance test methods. These test methods should also be continually reviewed as to their suitability for predicting the performance of the package in use.

Appearance General tolerances on appearance and printing are included. Again, accepted industry tolerances should be used where possible. For gravure printed roll stock, for example, a printing tolerance of ∓ 1/64 inch (∓ 0.4mm) may be specified. For glass containers a list of undesirable visual defects is included.

For printed packages reference is made to a complementary system of approved colour standards. These should be available for the type of material being specified. For example, colour standards for printed aluminium foil packages should also be printed on foil. The normal procedure in developing colour standards is to ask the supplier to submit samples from a production run which are typical of the light, standard and dark limits for a given colour. These are then submitted to the user companies' design department for approval or comments. Following approval the packaging department has the responsibility for numbering and distribution to supplier and to the quality control department.

Shipping Requirements for the packaging and identification of orders are covered in this section. Usually the supplier is responsible for the arrival of packaging materials in good condition at the user's plant. Care should also be taken to specify the outer packaging so that the packaging materials do not deteriorate in storage before use. Cartons and labels, for example, should be packed so as to prevent curl during storage.

The general specification is completed by a section for approved signatures. As general specifications are changed less frequently than detail specifications a long approval list is not a disadvantage. General specifications should be approved as follows.

Prepared by	_____	Packaging technician
Approved by	_____	Packaging manager
	_____	Engineering
	_____	Manufacturing
	_____	Purchasing
	_____	Quality control
	_____	Research

At the bottom left hand corner a space is provided for comments on reasons for change. This is for the information of people approving and receiving the amended specification. A small (c) notation can also be included in the margin opposite the altered section to further draw attention to the change.

All possible general information should be included in this specification to keep the detail specification as simple as possible.

Detail specification

These are prepared for each individual packaging item purchased. They consist of a single sheet. Each detail specification should be identified by a number. The first information to be determined before preparing a numbering system is the maximum number of packaging material items likely to be covered. In addition, as the system may need to be applied to a computer controlled inventory system, each number should identify each packaging material item individually. A typical system is as follows:

$$AA - BB - CC$$

1 **AA** represents the product group eg in a multi-product food company 01 refers to coffee; 02 desserts; 03 bakery products etc.
2 **BB** represents the packaging type and refers to the general specification number; 05 refers to cartons; 07 to glass containers etc.
3 **CC** is a sequential number identifying the individual package; eg 01 represents the 2oz size; 02 the 6oz size and 03 the 10oz size.

Examples of this system, again in a multi-product food company, are as follows:

01 − 09 − 03 − Instant coffee outercase, 10oz size
01 − 07 − 02 − Instant coffee glass jar, 6oz size
02 − 05 − 01 − Dessert carton, 2oz size

Form layout is similar to that used for the general specifications. Figure 6.2 shows a sample specification for a 6oz instant coffee glass container. Breakdown of the sections is as follows.

DETAIL SPECIFICATION

COMPANY NAME

PACKAGING MATERIAL SPECIFICATION

No: 01-07-02A   Type : Detail   Date:

Supercedes: 01-07-02   Glass containers   **Distribution:**   Manufacturing: Research
                                                   Engineering: Purchasing

6 oz instant coffee jar

1 CONSTRUCTION
    Colour: Flint
    Drawing no: No 69-T-7053
 (C) Weight: 11 50 oz $\mp$ 0.50 oz
    Capacity: 23 23 oz $\mp$ 0.312 oz (Brimful)
    Finish: 70 mm cap (TW)
    Coating: Hot and cold end coatings. Formula as approved by company.
    Overall height: 6.875" $\mp$ 0.047"
    Maximum diameter: 3.185" $\mp$ 0.062"
    Label diameter: 2.935" $\mp$ 0.062"

2 PERFORMANCE: To run without difficulty on properly adjusted packaging equipment at speeds of up to 200 per minute

3 APPEARANCE: As approved sample: To conform with General Specification No G.07.

4 PACKAGING AND SHIPPING: Jars to be packed in cases supplied by Company. Top flaps relative to printing to be glue sealed. Partitions inserted. Jars to be packed neck down. Pallet loads to be identified with production date, size and order number.

5 GENERAL   To conform in all aspects with General Specification No G-07.

REASON FOR CHANGE   APPROVALS
                                           PACKAGING
   Jar weight reduced from
   12 oz to 11.50 oz.   PURCHASING

Figure 6.2 Detail specification

Construction   This section gives details of dimensions, materials, style and basis weight. Individual tolerances are also specified where they are not covered in the general specifications. In the example in Figure 6.2 tolerances on glass containers for weight, dimensions and capacity vary, depending on the container size and hence are stated on the detail rather than general specifications. Reference is made where applicable to the standard drawing which should accompany the detail specification. More elaborate specifications sometimes include a sketch of the package component in this section.

Performance   Packing machine requirements are stated. In some cases this will involve only a statement that the material will perform at a given speed on a properly adjusted packaging machine. However, in more complex cases, as for example on form/fill/seal equipment, operating conditions such as temperature, pressure, dwell time will be specified. Limits on performance or laboratory tests are also specified here together with any performance tests not included in the general specifications. Examples of these are specific drop tests for plastic containers and compression test limits for corrugated outercases.

Appearance   This section states the type of printing and number of colours on the individual item. Full details of the design can be included here together with an attached photograph of the artwork. Standard drawings accompanying the detail specification should specify permissible print areas. In user companies with continually changing designs this section simply refers to the latest purchase order for design details. Colour standard numbers are specified for all colours mentioned.

Shipping   Individual shipping requirements as outlined by the receiving plant are stated. In the example of the instant coffee container (Figure 6.2) this includes the method of packing of the jars as they are received in the outercase.

General   Reference is made to the fact that the item should conform in all aspects with the relevant general specification. The reason for change, in revised specifications, is noted on the bottom left hand corner. The detail specification is completed by approvals signatures in the bottom right hand corner.

Setting-up and operating the system

Setting-up the system

The first step in setting up a new specification system of the type outlined in the previous section is to prepare drafts of general specifications for all the packaging groups. These drafts can be prepared only after a detailed study of processing and packaging conditions at the user's and supplier's factories. The packaging engineer, in the process of drawing up specifications, should discuss packaging needs and problems with all interested departments in his organisation. Shipping requirements, for example, can only be finalised after discussions with warehousing and materials handling personnel, while accountants may require certain information to be incorporated on the outside of packaging loads for assistance during stocktaking.

The draft specification is then discussed with the packaging supplier to see how it fits in with his normal operation. Any point raised by the new specification which makes the operation more complex and results in a cost increase for the packaging item should be assessed relative to its value to the user company. Once draft specifications have been agreed they are approved and issued to all suppliers and to interested departments within the user company.

Work can then commence on the detail specifications. Often the best method of selecting the required data is to request the information from the current suppliers. Machine performance requirements are obtained from the user company's own manufacturing and engineering departments. The numbering system can be compiled while waiting for this information.

Operating the system

The main objective once the system is complete is to keep it up to date with the minimum of effort. This is relatively simple because changes usually only involve re-issuing the one-page detail specification. It is important that the number of approvals on the detail specification be kept to a minimum. If possible only the packaging/purchasing departments should approve these specifications. Other interested departments such as research, manufacturing, engineering and marketing can be consulted where necessary and included on the distribution list. A long list of approvals can only increase the time necessary to issue a specification and hence reduce the efficiency of the system.

Both general and detail specifications are sent to all suppliers and reference made to the specification numbers on all purchase orders. Copies of blank specification forms are circulated to all departments on the distribution list to fill in when requesting changes in material specifications. Detail specifications which are superseded are identified by adding an alphabetical index after the CC number. A 'c' in brackets is included in the margin to indicate the change being made. For items under test, 'test specifications' can be issued. Coloured paper can be used to distinguish them from regular specifications.

Specifications as a management tool

The first section of this chapter described how the packaging specification gives vital help and information to many departments. It is also the main source of information for the packaging department. It enables the packaging engineer or chemist to keep a close check on all operations in which packaging materials function. The identification which a packaging material specification system gives to each packaging item can also be used as part of computer-regulated inventory control and purchasing systems. In this way the vast amount of information contained in any good packaging specification becomes available to other members of the management team and can be used to help solve other problems concerning stock control, determination of lead times and packaging material inventories.

The first objective of any function with responsibility for the technical aspects of packaging should be to have an efficient packaging material specification system. The actual type will vary from company to company depending on individual requirements. Close consultation is necessary during the preparation and maintenance of the specification system with all inter-company personnel concerned with packaging and with the outside material suppliers. An effective user/supplier relationship is necessary to achieve packaging material specifications which give the required performance and appearance at the right price. The guidelines expressed in this chapter are intended to assist the manager in the selection and development of a suitable packaging material specification system.

7

Control of Packaging Quality

The rapid growth of packaging in both the consumer goods and industrial products sectors has significantly influenced the methods used to control packaging quality. As already mentioned, in today's competitive supermarket-type markets there is a big demand for high quality pack appearance. In addition to this, a large proportion of packaging materials and containers are now handled by high-speed, semi or fully automatic packaging equipment. This equipment requires consistent packaging quality in order to function efficiently. These factors have led to a dual demand for high quality in areas of machine operation and pack appearance.

The packaging supplier, who originally was required to control appearance only, has had to raise his standards in this area and become familiar with the quality requirements of a wide range of packaging equipment. He will often find it difficult to relate his quality to that of the pack performance on the machine, whereas the user will readily be able to measure it in terms of output or machine efficiency.

The wording of the title of this chapter has been deliberately chosen to avoid the words 'Quality Control'. The average person, on reading these words, usually immediately thinks of inspectors in white uniforms, inspecting, sampling and often rejecting products, with resultant shut-down of processing or packing operations. In fact this type of quality control department does not directly control quality, it simply measures it. The actual control is inherent in the operation and is exercised by the machine operator or his supervisor, or by the controlling program in computer-monitored processes.

This chapter will deal with the basic principles of quality and with the methods of measuring it. The possible approaches which the packaging supplier and the packaging user can take in this area will also be discussed.

Quality and measurement

It was earlier described (in Chapter Six), that quality is built into a product or pack at the time of its manufacture. 'n this section quality and the methods of measuring it are defined.

What is quality?

There is a need to know not only what quality is, but also what aspects of quality are concerned in the packaging process. Quality is a difficult word to describe. The Oxford Dictionary defines it as 'degree of excellence'. The best way to imagine quality is to compare it to perfection. Manufacturers very rarely, if ever, produce perfect products. Quality is best thought of as the gap between what has been made and perfection. Quality is measured by the size of the gap, ie the smaller the gap the better the quality. Quality, is unlike perfection, in that it is a down to earth, measurable entity.

The two aspects of quality which mainly concern this discussion are, firstly the level of quality, (ie what is the size of the gap) and secondly, what degree of consistency is attained around that level. For example, when measuring the degree of whiteness of a paper or board on a scale, 100 may represent perfection while 90 is set as a quality standard. It is also necessary to ensure that the spread of results was consistent around this figure, (ie evenly spread within a range of plus or minus one, or five or ten), depending on the requirements. Both the target standard and the degree of consistency must be directly related to the requirements. Too low a standard results in unsatisfactory packages; too high a standard results in an over expensive package.

Specifying quality

Quality standards vary widely from one package user to another even for the same type of packaging item. For example, in the case of a plastic container the main concern of a toiletry products manufacturer will be with pack appearance. The household products man, however, will probably be more concerned with automatic filling at high line speeds. This difference in attitude will be reflected in the standards set. Standards can also vary between user companies within the same product area, again depending on appearance and manufacturing requirements.

However, for all packaging items, whatever the requirements, the quality cannot be controlled unless standards exist against which production samples can be compared. There are four main methods which can be used to set these standards. These are usually used in combination to form the major part of the packaging specification (see Chapter Six).

Physical measurements   These form the basis of the construction section of the packaging specification. The types of measurements are dimensions, weight, wall thickness, capacity, etc. These are stated together with the allowable tolerances. The methods for measuring this type of information are usually generally accepted and not included on the specification. Often standards of this form are accompanied by an official drawing giving most of the details in visual form. For a glass container, for example, the drawing would show a general profile of the container and detail all the important dimensions and their tolerances. Container weight and capacity would also be shown, again with associated tolerances. The neck finish would be marked, together with the type of thread and details of the necessary closure. For flexible packaging the drawing can be used to prepare a scale copy on a transparent plastic sheet. This can then be used as an overlay when checking either incoming packaging items or artwork, for size.

Standard tests   In this case the packaging materials are tested in a prescribed manner and the results must fall within specified tolerances. This type of test is usually included in the performance section of the specification. Typical tests in this category are:

1 Mullen test to measure burst strength of board.
2 Impact tests for glass or plastic containers.
3 'Peel strength' evaluations on flexible laminates to measure bond strength.
4 Dynamic compression tests on corrugated outercases.
5 Print rub resistance tests on cartons and labels.

With this type of standard it is important that the test methods concerned are specified concisely. Care must be taken that the test conditions used by the supplier and the user are identical.

Limit standards A limit standard defines the two extremes between which the packaging materials are acceptable. The main use of this type of standard is in the area of pack appearance. For printed packaging, for example, colour standards are agreed between the supplier and the user. These show a sample of the standard target for a colour or complete design and also examples of acceptable variation on the light and dark sides of the standard.

Acceptable quality levels For any package type there are a large number of potential faults or defects. These may vary from serious, and definitely unacceptable faults, to minor ones, which may be tolerated providing there are not too many present. If, for example, a metal can is considered which is printed by lithography, there are three possible defects, which can be classified as follows:

1 Critical : leaking side seams
2 Major : poor printing in main decoration area
3 Minor : scratches near bottom or top

A production run of cans, some of which had leaking side seams, ie critical defect, would definitely be unacceptable, whereas the acceptability of another run containing major and minor defects would depend on the number present. Obviously a greater number of minor defects than major ones, is allowable.

Sampling

Most activities concerned with the measurement of quality are also concerned with sampling. At first sight it might seem that the safest way to measure quality accurately is to look at every item ie 100 per cent inspection. There are two main reasons why this is not so:

1 Although costly this method is not infallible and there is still a chance that a defective pack will slip past.
2 The methods of quality measurement often involve destructive test methods.

The principle of sampling is that a sample is taken from a batch of items, in such a way that the quality of the sample is representative of the quality of the batch from which it was taken. When assessing the quality of a given batch by means of measuring a sample, there obviously exists a big chance that the result will not be typical. This can be reduced by statistical or random sampling. In this way each container has an equal chance of inclusion in the sample. The chance can, of course, be further reduced by increasing the size of the sample, but it must be remembered that the larger the sample the greater the cost of the measurement process.

There are many published tables of statistical sampling plans, as they are called. The original tables were developed by Dodge and Romig and resulted from work pioneered by the telephone industry in the USA. Most of these tables are issued by government organisations, who, because of their large and varied purchasing budgets have a continual need for effective sampling. A typical example is the Ministry of Defence specification DEF 131A — Sampling Procedures and Tables for Inspection by Attributes — HMSO London. These tables give a series of sample sizes and associated accept/reject numbers for a range of acceptable quality levels. They also contain 'Operating Characteristic' graphs which indicate the probability of any defective items not being detected by any given plan.

The two main points of statistical sampling plans are:

1 The nature of the situation to which they are to be applied must be known, ie we must have an idea what type and level of defect item is to be expected. In statistical jargon this is known as the 'Expectation'.
2 No sampling scheme is completely infallible. The limitations and

potential usefulness of the system being used must be recognised and accepted.

The next step in the quality measurement process is to examine and measure the sample which has been chosen.

Variability

The first point to understand when considering the measurement of a sample is that no two items are the same. This applies whether we are considering cans, cartons, aerosols or even elephants. Any given two items or a number of items will differ in some aspect. In using statistical methods of quality measurement we apply the knowledge obtained from the differences or variability in the sample to the whole batch (or population as it is often called) from which the sample was chosen. In some instances it is literally a case of ending up 'knowing everything about nothing', for often the measuring involves the destruction of the sample. A typical example of this is the measurement of polyethylene bag strength. The sample bags (say four) are subjected to destructive tests to measure the strength of the side seals. However in these cases the information obtained enables us to predict, with a degree of certainty, the measurements of the main population. The main point to remember about the sample is that it contains all the available information and that we must extract from it as much as possible. It should also be remembered that there is some chance that the sample will contain a freak result.

There are two characteristic patterns of variability which significantly effect quality measurement:

1 The Normal, or Gaussian distribution. This states that there is one result or value which is more frequently encountered than any other and that while other values exist on either side of this, their frequency diminishes as one moves away from the most common value. A good illustration of this is the height of men. Figure 7.1 shows a graph of frequency of occurrence against height. There are lots of men of 'average' height. There are as many who are just above average height and the same number who are just below average height. When moving away from average height it is found that there are fewer and fewer cases until the extremes are reached; the dwarfs

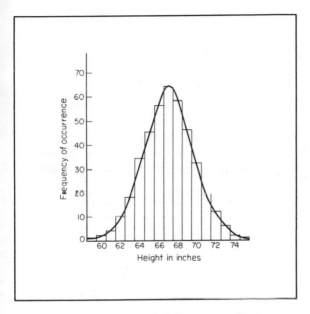

Figure 7.1 Frequency in height-groups of men

at one end and the giants at the other end.

2 The distribution of unlikely results (the Poisson distribution). This says, in effect, that unlikely results can only occur infrequently. For example, in a card game if you are dealt four aces it is unusual but not impossible. If, however, it happens more than once, say three or four times, then either the cards or the dealer are suspect.

Most manufacturing operations are found to conform to one or other of these distributions. Other forms of behaviour are possible, however, and it is always wise to check, before using tables based on these distributions.

When measuring quality, it is normal to extract two facts from the sample; the *average* and the *range*. The *average* is the number obtained by totalling all the individual results and dividing by the number of items in the sample. Care must be taken with averages for often they do not exist as such. Consider the case, for example, of the number of children per house in a sample of four homes. If the measurement is as follows:

> First house : 2 children
> Second house : 0 children
> Third house : 4 children
> Fourth house : 1 child

The total is seven children and the average 1¾. The average does not exist but in this case it is useful in compiling statistics.

The *range* is the difference between the biggest and the smallest measurement. The *range* tells us about the minute-to-minute variability.

Consider now a packaging example which illustrates the case of the *average* and *range*.

Suppose that on a machine a dimension must be held between 0.075 inch and 0.085 inch. A tolerance of 0.010 inch (or variability) has been allowed. Consider that the following readings were obtained from a sample.

Example A Specification. Minimum 0.075 inch — maximum 0.085 inch.

Specimen number	Reading
1	0.070 inch
2	0.090 inch
3	0.086 inch
4	0.074 inch
Total	0.320"

Sample average $= \dfrac{0.320}{4}$ inch $= 0.080$ inch

Sample range = 0.090 inch — 0.070 inch = 0.020 inch.

In this case the average is spot-on specification, but every reading is outside the limits. The average is right but there is too much variability. Now consider if the results had been as follows:

Example B

Specimen number		Reading
1		0.074 inch
2		0.073 inch
3		0.073 inch
4		0.072 inch
	Total	0.292 inch

In this sample there is very little variability (a *range* of 0.002 inch compared to 0.020 inch) but the average (0.073 inch) is just outside the specification (0.075 inch - 0.085 inch).

These examples have done more than just carry out measurements; they have shown a lot about the way the machine is behaving. In example 'A' the machine was jumping all over the place (*range* 0.020 inch). In this case there is a high probability of a machine fault. Example 'B' was altogether different, the readings were much more consistent (*range* of 0.002 inch) and all that was wrong was the machine setting.

It can be seen from this that a measurement of variability tells us two things, first, the instantaneous variation between successive items and second, the long-term variation or drift of the operation. When measuring variability two classifications are used, namely, variables and attributes.

Variables   These are measurable characteristics which can be evaluated on a continuous scale as, for example, weight, length, volume etc. The majority of variables which are dealt with conform to a Normal or Gaussian distributions.

Attributes   This covers subjective assessments by the operator or person responsible for quality measurement. The objects are assessed visually and classed on a discrete scale. A typical attribute is the measurement of finished pack appearance rated against a scale. Attributes are in fact only a short way of classifying the many influences which produce a result (for example, pack appearance). Each individual item could be classified as a variable if it could be isolated.

These two classifications are sufficient to measure variability. A characteristic can be measured (variables) or it can be judged (attributes), by rating or counting.

Having discussed the meaning of variability and the methods used to measure it, the following paragraphs consider its effect in day-to-day quality measurement.

Control charts

One of the main responsibilities of the quality measurement section is to advise the production management when their results show that the equipment is producing off specification materials or if the general drift is in that direction. It is important that this is only done when production conditions are in fact out of line and not because a freak result is detected. If the average is in the right place and the variation is under control, the production can continue. Control charts are used to give a measure of 'when to leave well alone' and 'when to do something' to the machine. There are two types of control chart, one dealing with variables, the other dealing with attributes.

Variables control chart   Averages are used on these charts, rather than individual results, because they tend to damp down the violent fluctuations of individual results and give an easier to read picture of the general pattern of results. The averages plotted are the individual averages for each sample (eg $n = 4$). It must be remembered, however, that the variability of averages is a lot less than that of individual results. This is taken into account when drawing in the control lines on the chart. The control lines for averages being closer together than those for individual results.

Figure 7.2 shows a typical control chart. Considering the average chart first it is seen that limits on this chart are closer together than those on the specification because it deals with averages instead of individual results.

When the averages fall within the control limits the process is behaving normally. When a result goes outside either of the limits another sample is taken to check whether the result is a freak or not. The result will give either:

1   Another result outside the control limits (as B on Figure 7.2), or
2   A result which is now within the control limits (as A on Figure 7.2).

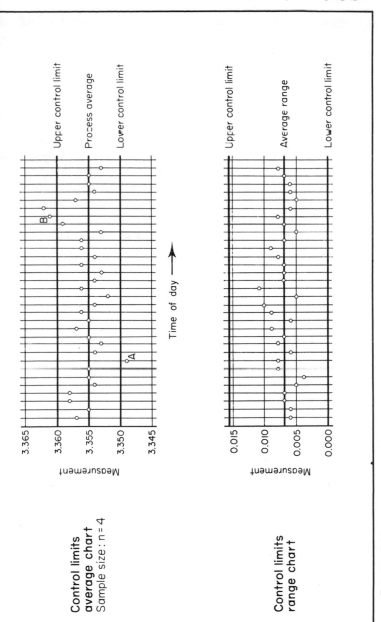

Figure 7.2 Variables control chart

If A is found, the extreme result was a freak and we can carry on as usual. But if B is found, something is wrong and action is taken.

This type of control chart helps to indicate the need to take action before trouble is reached. If the dots keep away from the control limits the situation is good; the warning sign occurs when the general drift of results is towards the limits.

The second chart shown in Figure 7.2 ie the 'Range Chart', gives us a measure of the variation at the moment, ie between largest and smallest item in our sample. To get a true picture, always look at the average and range charts together. That is why they are always printed on the same piece of paper. Control limits and the average line are also used on the range chart. If there is a large range within one sample and a small range within the next sample and this repeats itself it means that the process is jumping about. Either the components are varying or the machine needs attention.

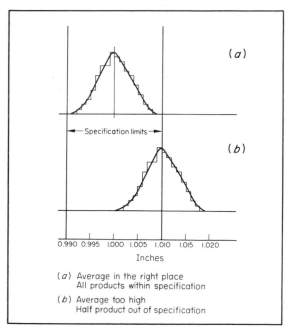

Figure 7.3 Spread of individual results

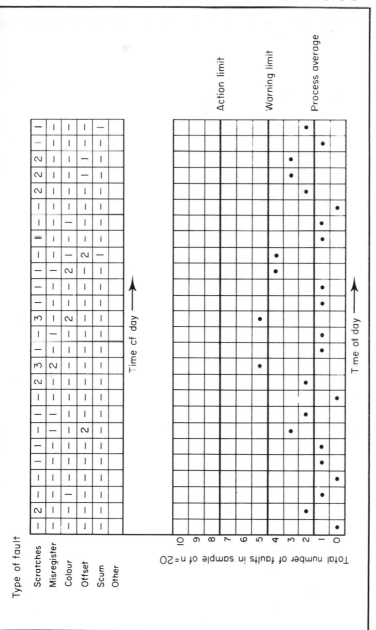

Figure 7.4 Attribute chart

Difficulties often occur not when the process will not keep to the tolerances but when the process average is in the wrong place, ie different to the specification average. Figure 7.3 illustrates a typical example of this occurrence.

Attribute control chart   The attribute control chart is similar to the variables chart in that limits are incorporated, based on either the acceptable quality levels set down in the user's specification or by in-plant standards. Often the in-plant standards are more severe than the user's specification, this being done to ensure compliance by including a safety margin. Usually a larger number of individual items per sample are required for the attributes chart compared with those required for the variables chart. This is not critical in cost terms, as acceptable samples can usually be returned to the production operation. Figure 7.4 shows a typical attribute chart for a printed tinplate container. The defects are shown at the top of the chart and the measurements down below. Control charts such as these are particularly useful to the machine operator who can immediately see the problem areas.

This concludes the theoretical aspects of quality measurement. The following paragraphs consider the methods used to control quality in the supplier and user organisations.

Packaging supplier quality control

The packaging supplier's main objective is to produce and deliver items to the package user, to a given specification, at the same time ensuring that his operation is profitable. To do this he must not only be able to measure quality but also to effectively control it.

Total quality control

It is now generally accepted that to control product quality effectively the control must be exercised right through the product life; starting from the design stage, through the production process, up to the point when the product is used. This approach of total quality control forms the basis of the 'Zero Defect' method used successfully in the USA in the aerospace industry. Application of this approach requires that checks and measurements are taken at each stage of the operation and compared against the required standards. The information from these multi-stage inspections is fed into a central control, usually the quality control department, and corrective action taken when any trends showing the possibility of defective quality are detected.

The aim of this approach is to ensure that everything is right, from the beginning to the end of the operation. The target is not to find and eliminate defective containers but to stop them occurring altogether. When using this approach it is emphasised that everybody concerned with the operation, not just the quality control department or the management, is responsible for ensuring good quality. This is particularly true of the machine operators who are encouraged to take an interest in the quality as well as the quantity of the containers which they are producing. This is done by explaining in understandable terms to the operatives what the aims and objectives of the programme are. Often incentive bonuses, based on the finished quality reports and complaints records are paid to these operatives. No total quality concept operation can survive without stress being placed on the philosophy that quality is everyone's business.

One of the most important areas in the control of product quality occurs at the design stage. At this part of the development people are often so concerned with the creative aspects of the design that other aspects, important to the overall quality, are overlooked. If, for example, a plastic container being developed for a liquid cooking oil product is considered, there are many quality-orientated factors which must be considered at the design stage. First of all care is taken that sharp corners are avoided in the basic shape. This helps to ensure that an even material distribution is obtained during the bottle-blowing operation. It also contributes to the container's resistance to both impact and environmental stress cracking. The latter is more likely to occur at sharp edges than at smooth radii, particularly for blow-moulded polyethylene containers. If storage tests have shown that the product has a tendency to exhibit panelling or wall collapse then the container can be designed to minimise the visual effect of this, for example, a flat oval shape is much better in this situation than a round cylindrical shape. The method of filling and, of course, application are also considered when designing the container. Finally, the methods adopted by the housewife to use, dispense and store the container are considered. It can be seen from this example that the plastic container designer is anticipating quality problems not only in the supplier's production operation, but also in the user's filling and distribution process and, last but not least, when the container is being used by the housewife.

Organisation

The quality control department is usually led by a manager who reports to senior management, usually independent of the production operation. It is now generally accepted that for an efficient operation the quality control function must be independent of the production responsibility. In the majority of packaging suppliers individual operations are run by a general manager, who has both the production manager and the quality control manager reporting to him. In the case of a disagreement between the two functions concerning quality he makes the final decision whether to ship, reselect or reject the items in question. With this type of organisation, while the quality control department does not have any direct control over the production operation (ie they cannot stop machines), neither does the production unit have the final say on overall material quality.

In the larger companies the department is split into two sectors: quality measurement and quality assurance. The quality measurement section is concerned with monitoring the day-to-day production to ensure compliance with specifications. They set up a measurement process for production and supply recording aids such as the control charts discussed previously in this chapter. The measurements themselves may be carried out either by trained personnel within the section or by the production operatives themselves. The quality assurance section consists of technicians who carry out laboratory tests as required and help to investigate machinery and process problems. They also investigate the reasons for complaint material sent back from the user company.

Responsibilities

The quality control department has three main areas of responsibility.

Raw material quality   The department has the responsibility for ensuring that all incoming raw materials are suitable for the production process, ie that they will, under normal conditions, produce satisfactory finished packaging materials. The normal procedure here is to ensure that the raw material supplier carries out all the necessary quality checks. Checks on incoming materials only being carried out in areas where problems are anticipated. Close contact is necessary between the quality control section and the supplier to carry out this type of quality assurance system.

Packaging material quality The second area is the assurance of producing packaging materials to the required quality at the lowest possible cost. The first part of this is covered by the supply of a quality measurement service to ensure that the production complies with the set specifications. Also a system must be set up for the elimination of defective material. The second part covers one of the most important functions of the department, to ensure that the goods are produced at maximum cost efficiency. This can be achieved by the identification of the defects which occur most commonly together with the reasons for their occurrence. Often these will be caused by either operator inefficiency or machinery faults, both of which can be reduced or eliminated. Elimination of defect categories in this way makes a direct contribution to overall profitability.

Customer satisfaction The quality control section is responsible for ensuring that the package user is satisfied with the quality of materials received. This responsibility involves close liaison with the user, and the supply of assistance when problems are being experienced. These problems may be caused by either the packaging materials or the machinery on which they run. The quality control technician helps the user to solve the problem and in the case of defective material follows up the complaint at his own plant.

Cost

As we have stressed earlier in this chapter, basically quality is controlled in the production operation. It is therefore possible to have a quality control programme without specific departmental responsibility. The onus for the control of quality in this case is borne by all the areas concerned with the operation, particularly the production operatives and managers. No company can exist without some form of quality control. The main aim of any quality control operation should be to eliminate as many defects as possible at the lowest possible cost.

Following are the two main cost areas involved in controlling quality.

Defective materials　There are several costs involved when defective materials are produced. First, there are those involved in isolating, removing and scrapping the defective items. Second, there are the costs associated with the machinery downtime and loss of production caused by the faulty items. The third cost item involved is the identification of the reason for the occurrence of the defects and the action taken to reduce or eliminate them.

Customer associated　Delivery of defective material to customers will almost certainly result in complaints and in some cases loss of business. The costs involved here include the time taken in dealing with complaints and the loss of goodwill (and possibly orders) from the customer.

From the above discussion it will be obvious that the application of a total quality control approach by the manager should succeed in either reducing to a minimum, or eliminating altogether the above extra costs.

Packaging user quality control

The quality control function within the packaging user's organisation can be broken down into two broad areas of responsibility:

1　Responsibility for the quality of incoming packaging materials.
2　Responsibility for the rest of the operation, involving packing or filling, storage, distribution and in-use.

Incoming packaging materials

The quality control section has the responsibility for ensuring that incoming packaging materials conform to the agreed specifications. This responsibility is broken down into two main areas, the first concerning pack appearance, the second concerning pack performance. The appearance requirements are that the design or colours of the pack should fall within the approved limits. It is necessary to make sure that there is not a gradual drift, with time, away from the standard. Comparisons of incoming materials, for pack appearance, with the standard should always be carried out under standard lighting conditions.

Pack performance is usually rated in terms of efficiency in running on the packaging line, or through the manual packing operation. Often, in problem areas, simulated test methods are developed which are typical of machine running conditions. These are carried out on incoming packaging materials to try and anticipate machinery problems and prevent costly machine downtime.

The methods used to check incoming packaging materials vary widely from one company to another, but can broadly be classified into three main sections.

Large inspection unit   The user in this case carries out some form of inspection on all the packaging materials received. The extent of the inspection can vary from 100 per cent to a statistically based sampling scheme. The 100 per cent inspection method is very costly, and still not a completely effective method. It may, however, be justified at times of crisis when extreme measures are necessary to keep production going. The acceptance sample scheme method is the one most used by companies, who for one reason or another, decide that they need to inspect all incoming items. This type of scheme describes the sampling technique, specifies the defect groups, together with their acceptable quality levels and lists the action procedures.

Random checks   The normal procedure in this situation is to employ an inspector to carry out random checks on incoming materials. This is done by taking a relatively large sample from a given delivery and examining it. The group of materials examined varies from day to day. For example, glass containers may be examined on the first day, metal containers on the second, etc. Often when using this method the attention will be focused more on the packaging material items which are most important to the overall operation. This method has the advantage that when problems arise there is a single individual, experienced in inspection, who can work full time in the area of concern. Generally this method works better than using a single inspector to try and check all incoming items.

Quality assurance programmes   This is rapidly becoming the most popular method, involving close co-operation between the supplier and user. Basically, the principle is that the user relies on the supplier to produce satisfactory materials and does not carry out any incoming checks. Many package users have worked on this basis for a long time simply relying completely on their suppliers. Modern quality assurance programmes go further than this as both parties concerned work together to control the quality. The user works with the supplier to ensure that all aspects of the production process are conducive to producing good quality products. This is done by the user visiting the supplier's plant and assessing not only the control facilities, but also the general manufacturing and process conditions together with plant house-keeping. The user then reports on this overall operation to the supplier, who in turn progresses any indicated action points. This method ensures that quality control effort is not duplicated and that both the user and the supplier are committed to producing quality materials.

Finished pack quality

In the same way that the packaging supplier must ensure that his materials comply with the specification so the user must ensure that his finished pack quality meets a given standard target. In the in-plant operation of the package user there are two main areas which are monitored by the quality control department:

1 The packaging material performance in the packing operation.
2 The measurement and control of finished pack quality.

Packing line operation   The main responsibility of the quality control section in this area is to investigate complaints of machine downtime or lost production, and to help determine the cause. This is often a very difficult area, for it is a complex problem separating the packaging machinery/material interface. Often packaging materials are blamed for poor machine performance and vice versa. It is necessary, therefore, for the quality control technician or inspector to become familiar with the machinery and be able to isolate not only packaging material complaints but also point out problems caused directly by the machinery.

Finished pack standards   These usually consist of two main types. There are standards governing the finished pack appearance and there are performance tests to ensure that the finished packed product will stand up to the storage, distribution and handling until the end of its life. The appearance standards are often in the form of attribute control charts specifying the limits for likely defects. Typical of the performance tests would be drop tests carried out on plastic containers to confirm that the impact resistance of the pack is adequate.

The quality control section or department in the operation is organised in a similar way to that of the packaging supplier, with the manager independent of the production operation. It is not usual to have a separate quality measurement section, normally all areas of responsibility are handled by the technicians or inspectors. Often the packaging quality control section is one unit of the overall plant quality control department.

In conclusion it should be stressed that there is nothing magical about quality control. The function of the quality control department is to assist other departments to achieve good quality. To this end, a total commitment is required by all concerned. This applies not only within a single company or industry but right through the chain of people involved in the packaging process, from the raw material supplier, through the packaging material producer to the packaging user.

8

Economics of Packaging

Economics can be defined as the science of production and distribution of resources. Packaging materials and containers are included under the resources heading. They concern not only the production and distribution but also the selling of resources (or products).

Earlier chapters described packaging as an integral part of any product and explained that it can affect the relative success at several stages of the product's life. For example, when considering a consumer product selling in a supermarket environment, the surface design or package shape can have a significant effect on sales. For an industrial product, however, the identification and physical protection provided by the pack are the critical factors. In each of these cases the cost efficiency of the pack is identified not only by its basic cost but also by its effective contribution to the whole operation.

The actual unit packaging material cost as a percentage of the total production cost varies according to the product type. For commodity food products such as sugar and flour, it is usually below 5 per cent. For other consumer products such as detergents, toilet soaps and biscuits the figure varies between 5 per cent and 10 per cent. In the areas of toiletries and cosmetic products the packaging cost is often equal to or greater than that of the product itself.

There is no direct relation between product and packaging price. For example, expensive items such as electrical appliances are often packaged at a cost of less than 5 per cent of the total production cost. The importance of packaging to the national economy can be measured either by the total amount spent per year, or by the *per capita* expenditure.

In the UK it is estimated that in 1971 the amount spent on packaging materials and containers exceeded £1,200 million, equivalent to approximately £24 per head of population. The corresponding figures for the USA are £9,000 million or £45 per head. Both these sets of figures are based on packaging material and container costs, before processing or packing, and would be at least doubled if the values were calculated at the point of sale.

Cost considerations play a major role in the work of all the people concerned with packaging operations. In this chapter we will deal with the aspects of packaging economics with which the manager is concerned in day-to-day operations. The areas where packaging cost and performance affects the total product cost will be discussed and methods of cost reduction in these areas will be studied. The chapter is completed by a section on the methods used to justify capital investment in the packaging area.

Elements of packaging cost

In Chapter One the breakdown of the areas of packaging cost was briefly discussed. It was stressed that costs associated with packaging are incurred right through any operation and that basic material or container costs should never be considered in isolation. There are many case histories in the packaging area where changes made on the basis of considering a packaging material price reduction only, have ultimately resulted in an overall increase in total product cost instead of the expected saving. In this section the elements of the total packaging cost will be considered in detail.

Unit pack cost

This is one of the easy-to-measure areas of cost as the basic material or container price is stated on the supplier's quotation. This quotation will be based on either the user's specification, for custom manufactured materials, or on the supplier's own specification for stock items. The actual unit price on the quotation will depend on the following factors.

Volume   For the majority of packaging material items the unit price will be directly affected by the quantity ordered, ie the larger the order the lower the price. This is particularly true for manufacturing operations involving large machinery such as glass container manufacture or gravure printing, where the start-up, or set-up, time in the supplier's operation is a lengthy process.

The quantity ordered will also often dictate the type of process or machinery to be used for a given order. For example, let us consider a flexible laminate to be printed with a four colour design. For orders of say, 100,000 impressions, the supplier would use flexographic printing. For larger orders of say, over a million impressions, the use of gravure printing would be economically justified. In this type of situation there is a break-even point above which gravure printing gives a lower unit price than flexographic. The exact location of this point depends on the specification of the material in question and also on the exact type of equipment used.

Methods of overpacking   The supplier's quotation is usually based on safe delivery of the materials to the user's plant. This he may do by using his own transport or by hiring contractor's vehicles. The type of overpackaging to be used should always be clearly agreed between the user and the supplier.

Often more elaborate overpackaging, than is necessary for the shipping operation, will be required to enable the user to store and handle the items before use. For example, in the case of five-litre plastic containers the cheapest method of delivery, from the supplier's viewpoint, is to pack them in polyethylene bags and then stockpile direct into the transport. Often it is worth an extra charge, for the user to have the containers delivered packed on pallets, which can be easily stored and handled in his plant. Often special overpacking is specified by the user to suit high-speed packaging equipment. Metal cans, for example, can be supplied packed on pallets, with layer pads, ready for automatic feeding onto the packing lines.

Freight charges   The unit price will be directly affected by the delivery instructions on the user's purchase order. Often user companies place annual contracts with their suppliers and call in quantities as required. In these situations, the minimum quantities in each delivery are agreed between the two parties. Normally the supplier will make an extra charge if certain minimum quantities, eg a container load, are not ordered, to cover his extra freight costs. The unit price is also, of course, affected by the distance of the user's plant from that of the supplier. In situations where returnable overpackaging is utilised (for example, pallets or corrugated containers), the supplier's delivery vehicles collect these materials following the regular order deliveries.

Development costs The costs involved in the development of new packaging items have been discussed fully in Chapter Four (Approach to Packaging Development — Execution). The usual procedure in the packaging industry is for the supplier to absorb the initial development costs and then spread them over several years' packaging material items. It is likely therefore that the unit pack price will include a figure to cover the development costs which have been incurred. An alternative procedure is for suppliers to include research and development costs in their overheads and spread them over all items sold.

Storage/handling

In the previous section we have discussed the effect that order volume can have on the unit pack price. However, the price advantage gained from large volume ordering has to be considered in comparison to two other factors. First, the amount of capital which will be employed in holding large packaging material inventories and second, the cost of labour and space involved in the storage and handling of the materials. One of the buyers' main responsibilities is to study the relationship between unit pack price and volume and to decide what level of inventories to maintain. He will be influenced in his decision by the costs involved in the handling and storage of the materials. For example, the storage costs for items with a large volume/weight ratio, such as glass or plastic containers, are high compared to say rolls of flexible packaging material. In some instances where large orders are placed, the supplier will agree to store the packaging items himself, ready for call-in by the user when they are required. In this situation the extra storage costs incurred by the supplier will probably be included in the unit pack price.

The other costs incurred in this area are those caused by damage to materials or containers during storage and handling. These may take the form of breakage of glass containers, resulting in a direct loss or of damage to cartons or flexible packaging which will adversely affect the material's performance on the packaging machinery.

Production operation

The production of packaging materials to close tolerances, suitable for running on high-speed automatic equipment, makes a vital contribution to overall efficiency in today's high speed operations.

Cost areas   There are three main areas of cost in the packing or filling operations; materials, labour and overheads. Several different accounting methods are used to allocate these costs. One of the most commonly used is to consider them in two sections; fixed and variable.

The fixed costs for a given packaging line cover the machinery depreciation, the direct labour employed on the line and a proportion of the total overheads. The variable costs cover the raw materials and packaging used.

Fixed costs can be lowered by more efficient equipment utilisation or by a reduction in labour costs or overheads. Variable costs are reduced by savings in raw materials or packaging.

Machine efficiency   The overall efficiency of a packaging machine, or line, is judged by the cost of the finished product coming off the line. This is influenced not only by the machine speed and the labour involved but also by the unit pack cost of the materials used. The space occupied by the machine is another influencing factor. The capital cost of the machine also influences the efficiency, as a proportion of this is included in the overheads as depreciation.

Developments in both packaging materials and equipment have significantly reduced costs in these areas for many products in recent years.

Some examples follow:

1 Outercase forming systems handle flat blanks instead of glue-sealed outercases; reduction in unit pack cost.

2 In plant thermoforming of plastic containers; reduction in unit pack cost.
3 Development of high-speed liquid filling lines for aerosols; improved output.
4 Methods of hot-air sealing for wax-coated cartons; improved output.
5 Use of hot-melt adhesive systems; reduction in space.

Many of these improvements have been made possible by a 'systems' approach, whereby the packaging material supplier and the equipment manufacturer work closely together to achieve better performance. Many suppliers now market both machinery and the materials to run on it. In some instances the equipment is leased to the user.

From the above it will be obvious that changes in packaging material specifications can significantly effect machine efficiency In Chapter Four the importance of carrying out proper machine trials before changing specifications was pointed out. This point is stressed as it is vital to ensure that new or improved materials will function efficiently on equipment prior to the start-up. One practical point to remember, when calculating machinery efficiency for cost evaluations, is that for the majority of equipment some downtime will be necessary for reasons such as breaks, changeover of packaging material, etc.

Labour costs Not all packaging operations are automatic or even semi-automatic. There are still many operations which for reasons of either complexity or low output are completely manual. The manager's aim in this situation is to achieve maximum output per packer or operator. The main secret of success here is the efficient planning and layout of the packing operation. Where possible, conveyors should be used to pace the line.

Effective supervision also plays a big part in the achievement of high outputs from this type of operation. Often, incentive bonuses, based on output, are used as a method of raising production levels.

Packaging losses There are many potential areas of packaging material losses or 'shrinkage' in the production operation. Some of these will be unavoidable, for example, materials which are sampled by the quality measurement section and tested to destruction. It has already been mentioned in the section on storage and handling costs, that losses can be incurred by damage during this period. Contamination can also occur here if the materials are not adequately overpackaged.

The other two main potential loss areas are both concerned with the packaging operations. First, there is a potential loss during machinery start-ups, whether it be at the beginning of a shift, after a break, or after a size changeover. Second, there are the losses caused when machinery problems are experienced and off-standard packs are produced.

The losses discussed above are small if considered individually but can involve significant amounts of money when added up over a period of say a year. The normal accounting procedure, when dealing with packaging materials is to allow for a standard 'shrinkage' factor. This will vary from material to material, for example, for corrugated outer-cases it is usually less than 1 per cent, while for flexible packaging materials it can be as high as 5 per cent, even for an efficient operation.

Warehousing

After the production operation, the packed product is normally shipped to the user's warehouse for storage prior to shipment. This then becomes another element in the total product cost. In some cases special storage conditions are required which may involve control of temperature and/ or humidity. Frozen foods, for example, must be kept under deep freeze conditions at all stages of storage and distribution.

In addition long storage periods are often necessary for certain frozen fruits and vegetables which have short harvesting periods. In these situations the storage costs for the packed product are high. One solution to this problem is to blast freeze the fruit or vegetables after harvesting and store in bulk before packing as required.

Package shape has an influence on storage costs because it affects the utilisation of space. This is obviously important in the case of frozen foods, where the cost of low temperature storage is high. The packaging manager, when considering outercase design has to take the space utilisation factor into account along with the requirements for stability, economy, surface design, etc. When changing packaging speci-fications involves a change from one material type to another eg glass to plastic, carton to pouch, the effect of the change on package volume should be considered. Often a change of this nature can significantly reduce the storage space occupied by a pack, and hence bring down the total packaging cost.

Package strength is another factor which can affect the costs of storage. In today's modern single storey warehouses, products can be stored in two ways, either in racks several cases high, or dead stacked (with or without pallets), up to the roof. The latter method requires that, the outer pack and its contents, stand up to large stacking heights. If the more expensive method of rack storage is used, then the outer-case strength is not so important. In some situations however, it is more economical, for fragile products, to use rack storage rather than to specify expensive outerpackaging.

Distribution

The operations involved in moving the product from the user's warehouse may involve the use of several different forms of transport. In the UK for example, road, rail, or a combination of the two, are the main methods employed. For export shipment, the transportation may be by sea or air. The performance and physical properties of the package can have a significant effect in this area.

Transport costs   These are governed by either the finished pack weight or volume. They are also, of course, dependent on the shipping distance and the value of the item being handled. Air transport is a typical example where the total weight is the main factor affecting the cost for a given journey. This can also, however, occur with road transport where the product has a high weight to volume ratio and the transport is loaded to its maximum permitted weight, before it is completely full. In the majority of cases of shipment by road and rail, the package volume is the main element considered when calculating charges.

Package weight and dimensions can, therefore, affect costs in this area. For normal road and rail transport a change in pack volume or 'shipping cube' as it is called, often effects the distribution costs. An example of this is the replacement of outercases for the packaging of bags of coffee by paper baler bags. The new package, not only has a lower unit price, but also by nature of its lower volume, results in a lower distribution cost. For air transport, a reduction in package weight can have a major effect on cost.

One good illustration of this comes from the airlines themselves who achieved a major cost reduction by changing to low-weight PVC containers for spirit miniatures to replace the conventional glass containers.

Protective packaging The level of losses, due to either damage or pilferage, which are experienced during distribution make an important contribution to the total product cost. These are significantly affected by the protective properties of the package used. As explained in Chapter Four ('Approach to Packaging Development – Execution'), identification of the typical hazards likely to be encountered in the distribution operation, is one of the actions carried out in the packaging development programme. However, in this type of work excessive conditions and unexpected hazards are not normally guarded against. The reason for this is that it is generally not economical to produce packaging capable of protection against all possible conditions. This means in practice that a certain loss level in the distribution system is accepted for the majority of products. The packaging manager has the responsibility for deciding what this level is, and also for investigating the effects of small changes in packaging cost on this level.

Figure 8.1 shows graphically the relation between packaging cost

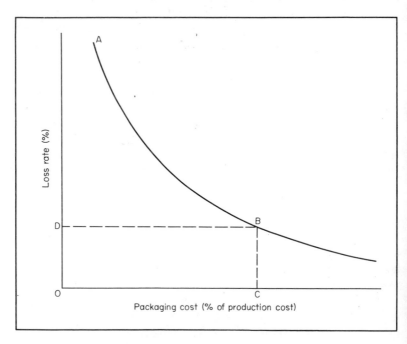

Figure 8.1 Graph of packaging cost v. loss by damage

and loss through damage, the packaging cost being expressed as a percentage of the production cost, and the loss as a percentage of the total consignment. It can be seen that starting from the point 'A', the loss rate decreases rapidly as the packaging cost increases. At the point 'B' the situation is seen where an increase in packaging cost results in only a small change in the loss rate. This is a typical situation for many packed products, 'C' is the packaging cost and 'D' represents the average percentage loss. The relative position of the curve in relation to the axis will depend both on the nature of the product and on the type of packaging employed.

In deciding on the packaging cost, the manager must consider the effect of the potential loss on his customer goodwill. Also if improved systems are developed for the packaging of a product (ie the curve in Figure 8.1 moves closer to the axis), the manager must decide whether to maintain the same packaging cost and reduce the loss rate, or to reduce the packaging cost and keep the loss rate constant. Also, he will consider the effect on his customers when taking this decision. The main points in this section are that the manager should monitor the losses in the distribution system and be aware of the effect on these losses of changes in packaging cost.

Selling operation

The first part of this chapter dealt with the packaging cost areas from the initial concept or design stage through to the distribution operation. As mentioned earlier the package can often directly influence the sales of a product, particularly in a supermarket-type environment. Often a component of the packaging, for example, a display carton, is specifically designed for this purpose alone. The effect of the package on sales can only be measured by a market or store test. This aspect of packaging economics is fully covered in Chapter Ten ('Packaging as a Marketing Tool').

Packaging cost reduction

In the first section of this chapter the areas in which packaging affects the total product cost were considered and the point was made that all the elements of the total operation must be considered to obtain the true packaging cost. The same applies when considering packaging cost reductions. The unit pack cost is only one of the factors which make up the total packaging cost.

Often it is possible to obtain an overall cost reduction by an increase in the unit pack cost. One example of this is the packaging of a chocolate-coated confectionery product, where a change from a paper wrapper to a more expensive paper/foil/polyethylene protective laminate, enabled the manufacturer to reduce the weight of the chocolate coating, resulting in an overall cost reduction. In this case the extra chocolate had been used to protect the inner product core, a role taken over by the new package.

The other important point about cost reduction exercises is that one of the objectives must be to achieve the savings without an effective change in quality or reliability. For example, when reducing the unit pack price by, say, a change in material supplier, it is essential to ensure that the new supplier can deliver the materials, not only at a lower price, but also at the same quality level.

Cost reduction programmes

The preparation and implementation of a packaging cost reduction programme is one of the main responsibilities of the packaging function. It normally comprises a large part of the total development programme and the results obtained often, by themselves, justify the existence of the packaging department.

Figure 8.2 shows a typical development programme for a multi-product food company. This includes both new product developments and cost reduction projects. For the latter the estimated saving is stated together with the costs involved in achieving it. Note also that for the new product development items, code names, in this example colours, are used for security reasons. Project Blue for example may involve the development of packaging for a new dessert product, while Project Yellow will concern a new confectionery item.

As mentioned earlier the packaging manager has the responsibility for the preparation and implementation of the cost reduction programme. In doing this he needs to study all the areas of packaging cost and identify potential cost saving possibilities. It is vital in this area that the packaging manager liaises effectively with all the people involved in the packaging operation to draw together all potential cost

COMPANY NAME

PACKAGING DEVELOPMENT PROGRAMME 1973

Project type	Objective	Timing	Estimated cost
1 NEW PRODUCTS			
Project blue	Develop packaging	National launch: May '73	£60,000
Project yellow	Develop packaging	Test market: August '73	£10,000
Project indigo	Develop packaging	Test market: September '73	£6,250
2 COST REDUCTIONS			
Margarine containers (£40,000/year)	Install in-plant thermoforming operation	July 1973	£65,000
Dessert cartons (£20,000/year)	Install cartonning system	November 1973	£15,600
Pet food outercases (£15,000/year)	Revise board specifications	June 1973	£5,800
Glass coffee containers (£10,000/year)	Reduce container weight	October 1973	£14,500
3 PACKAGE IMPROVEMENTS			
Pet food products	Introduce new label designs	May 1973	£3,000
Instant coffee	Develop new outercase size	June 1973	£540
Confectionary	Develop new pack size range	November 1973	£2,220
4 NON PROJECT			
Revise specifications	Introduce new computer based system	December 1973	£2,100
Purchase laboratory equipment	Improve testing facilities	November 1973	£3,500
		TOTAL	£188,510

Figure 8.2 Packaging development programme – 1973

157

reduction ideas. This applies not only to the departments within his own company but also to all his packaging material suppliers. Often it is helpful in this area if the packaging manager identifies the area of high packaging cost and sets objectives for the cost reduction effort. Having gathered together all the potential cost reduction ideas, the packaging manager then decides on the basis of the likely degree of success, and of the amount to be saved, which ones to include in his programme.

Areas of cost reduction

Reductions in packaging costs can be achieved by savings in any of the areas described under the 'Elements of Packaging Cost' in the first part of this chapter. Some examples of cost reductions have already been given under the various headings in this section.

In addition to the areas discussed, there is one other major factor which can affect the packaging cost, and this involves the product itself. Often changes are made to the product which change the degree of protection required from the package. Some examples, where a change in product formulation resulted in a lower packaging cost follow.

Toilet Soaps Traditionally toilet soaps have been packed in aluminium foil laminated wrappers, the packaging providing protection against perfume and moisture loss. Recently more stable soap formulations have been developed which enabled the use of cheaper plastic coated paper wrappers instead of the foil laminated materials.

Electrical appliances One of the reasons for significant cost reductions in this market sector has been the attention paid to packaging requirements at the product design stage. Small changes in product design have been found to reduce significantly the amount of cushioning required and hence result in a reduction in packaging cost.

Chemical products Corrosion of metal containers by liquid chemical products is one of the packaging problems encountered in this field. The normal procedure for corrosive products is to specify internally lacquered containers to prevent package deterioration. Recently many corrosion inhibitors have been developed, which when added to the chemical product, eliminate the need for lacquer lining. In these situations the cost of the inhibitor is often less than the premium involved in metal container lacquering.

The above examples illustrate how the product can affect the packaging requirements. Equally, changes in other parts of the operation can also affect the type of packaging required. A typical example of this in the distribution area is the growth of containerisation. This relatively new method of shipment, whereby goods are packed into a container which is shipped by either road, rail or sea direct to the ultimate customer, has significantly reduced the hazards to which packs are subjected during distribution. Many packaging users have found that they can reduce the performance requirements for the packaging of products to be shipped in this manner and hence reduce costs.

Value analysis techniques

'Value analysis', or 'value engineering' as it is sometimes called, is a modern management cost reduction technique which can be successfully applied to packaging operations. Basically value analysis works on the principle of assigning a function to every item involved in the operation and then studying alternative methods and costs of achieving that function. It can also be used as a method of achieving increased performance without an accompanying cost increase.

In the first part of this section the setting up of the cost reduction programme was discussed. Value analysis can be a valuable tool in this work. Many people argue that the normal type of logical cost reduction exercise and value analysis are synonymous. In many cases they are right. However, there is one basic distinction. Cost improvement projects often involve small changes in specification, for example, a change in board weight or type for a paperboard carton, a reduction of film weight for a flexible pouch, etc. whereas value analysis takes a much broader look at the whole packaging operation.

The objective of a value analysis approach is to obtain the maximum function at the lowest cost. Three basic steps are involved.

1 Define basic function of each component.
2 Propose alternative materials or methods to perform the same function.
3 Compare costs of current and alternative methods.

Application of this technique involves questioning of all the criteria used to develop the package. Often products are deliberately over-packaged, because at the development stage, the time limitations did not permit full testing. Value analysis in this situation helps bring down the packaging to its correct cost level. Often situations will be revealed where criteria which were applied at the development stage have now ceased to be relevant.

One example of this type of situation involved a change in distribution method for a packaged convenience food product. Originally the product was sold through distributors and was packed in outer-cases containing four inner cartons, each holding twelve individual packs. The purpose of the individual twelves cartons was for the distributor to use for shipment to his small customers. A change to direct selling, cutting out the distributor, meant that a single outer container became a satisfactory pack. This potential cost reduction was only revealed during a value analysis study.

The packaging buyer can also use value analysis to assess suppliers' quotations and to decide if the prices are reasonable. In this situation the buyer studies the operations involved in the production of the packaging item and also the raw materials used. He then assigns a value to each of these. When new manufacturing methods or new material specifications are proposed he is in a position to calculate the expected effect on the packaging price and compare it with the supplier's quotation.

In today's economic conditions where cost reduction is given the highest priority, the use of the formalised technique of value analysis can significantly help the packaging or purchasing manager to achieve his targets.

Justification of capital expenditure

The majority of packaging development projects involve expenditure in one or more, of the areas of materials, machinery and personnel. For cost reduction projects the expenditure is justified by the savings achieved. For improvements to pack appearance, or performance in use, the outlay is covered by a potential increase in sales. When dealing with the purchase of new packaging equipment large amounts of money are often involved, particularly for high speed automatic machines. There are three main methods of justifying expenditure of this nature.

Pay-back period

This method involves justifying the project on the basis of the length of time that the original capital outlay is at risk. For a cost saving project involving the purchase of a new machine, the cost of the machine divided by the annual amount saved would give the pay-back period. The shorter the period the more attractive the investment. For example, the pay-back period would be three years for the installation of a packaging system costing £9,000 which achieves an annual saving of £3,000.

This is a relatively simple method of operation and is adequate for many of the situations encountered with packaging operations, provided the limitations, stated below, are recognised. The standard procedure is for a company to set a maximum pay-back period, say three years, which all projects must achieve. The maximum pay-back period can be varied depending on the current financial situation. The following limitations should, however, be noted. Firstly, the method does not take inflation into account, ie it assumes that the value of money is constant throughout the project life. This is not too critical for projects with a short pay-back period. Secondly, it does not take into account the cash inflow which occurs after the pay-back period. This can be quite critical when using the method to compare projects, for it favours the project with the quickest return on investment regardless of the total proceeds over the whole project life.

Return on capital employed

This method attempts to take into account the total profits which accrue over the life of the project, independent of the rate at which they are earned. The profits are then measured as a percentage of the capital investment. If, for example, an investment for a packaging line with an estimated life of ten years costs £25,000 and the annual profit yield resulting from this has been calculated at £2,000 for the first three years, £3,000 for the next three years and £3,500 for the final four years then the calculation of rate of return is as follows:

Average capital employed   £12,500

Total profit over ten years   £29,000

Average annual profit   £ 2,900

$$\text{Rate of return} = \frac{£2,900}{£12,500} \times 100\% = 23.2\%$$

The average capital employed figure is obtained by taking the original investment of £25,000 and depreciating it over the life of the project. This gives a figure of £12,500 for the average amount employed over the whole project life.

This method does take into account the full life span of the project. It is useful for packaging projects which are based on marketing forecasts of increased sales where the annual amounts are incremental. However, this method also suffers from the disadvantage that inflation is ignored. In the example given above, inflation would have a significant effect on the rate of return and hence on the project justification. Despite this disadvantage it can be a useful method, particularly as it is based on standard accounting procedures. Normally, when using this system a company will set a target rate of return which all projects must attain.

Discounted cash flow

One factor of fundamental importance to investment decisions, which has not been taken into account in either of the two methods discussed, is the timing of profits and cash flows which result from an investment. A relatively new method of cost justification known as 'discounted cash flow' is based on the measurement of cash flows and their relation to current values. This method is now used by most major companies for the justification of capital investment programmes.

When preparing a justification for a project by this method, it is necessary to calculate the cash flows for each year of the project's life and then discount them back to their present value. This is illustrated by the example in Figure 8.3. This shows in (a) the cash flows for a packaging project justified on the basis of improving total sales. In the initiation year there is a negative cash flow accounted for by the purchase and installation of new equipment for the improved package. The cash inflows increase in the first years of the project's life reaching

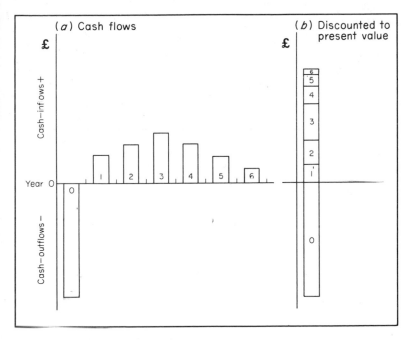

Figure 8.3 Discounted·cash flow

a peak in the third year, then dropping off to the end of the project's life in year six.

In (b) of Figure 8.3 the cash flows are discounted back to their current value. The discounting takes into account both inflation and the effects of compound interest. The discounting can be carried out in two ways either by using a standard discount rate for all projects or by calculating the discount rate which will bring the total cash flows down to a present day value equal to that of the total expenditure. Standard tables are available to assist with the discounting process. The calculations involved in using this method are however often complicated and it can only be justified for relatively large investments. The method has the advantage, however, that when the calculations are complete, the manager can see the actual cash flow over a period of time, relate it to current values and directly compare projects competing for investment resources.

In this chapter some aspects of packaging economics which affect the manager in his work have been discussed. The point that packaging is an integral part of the product and can influence the total cost at several stages of the operation has been described. Packaging is changing continually; not only are new materials being developed and new machines being introduced, but also new methods of storage, distribution and marketing are being pioneered. Any of these can affect the choice of package and hence the cost. Packaging personnel in today's business atmosphere, must continually monitor costs and look for potential improvements to ensure that efficiency is being maintained. Modern management methods such as 'value analysis' and 'discounted cash flow' are valuable aids in this task.

9

Packaging Material Trends

Material or container changes to improve efficiency in the areas of appearance, protection, function or cost, account for a large part of any packaging development programme. Previous chapters described examples of changes in packaging material specifications made necessary by the introduction of new or improved methods of packing or filling, warehousing, distribution and selling. In this chapter the current and predicted trends for the various packaging material and container groups will be outlined.

In the UK the statistics on the usage of packaging materials by industry are far from comprehensive. Up to 1965 there was no single reference document for this purpose. Since then PIRA have carried out statistical surveys for two periods (1965-1968 and 1967-1970), and these have greatly improved the situation. This contrasts with the USA where the individual packaging industry associations issue comprehensive figures and complete breakdowns of usage by packaging type are published annually in the Modern Packaging Encyclopaedia.

Figure 9.1 shows a table of estimated packaging material values, together with the percentage breakdowns for each group, for the UK in 1972. These figures are based on the PIRA survey by Rowena Mills, published in 1971, together with up-dated information from the industry associations, and government statistics. Tinplate containers and plastic items represent the two biggest individual groups. These figures contrast significantly with those for 1970 when tinplate containers were the leading group by a clear margin. The plastics heading for the purpose of this table covers the complete range of items used in the packaging field. If paper and board are taken together they comprise the largest material group, as in the majority of industrial countries. The other significant material group is glass containers, which despite intensive competition, is maintaining its share of the total market.

The main change in the overall positions over the last few years has been the growth of plastics at the expense of the metal, paper and board and glass sectors. The largest individual growth area, albeit from a relatively small base, has been that of aerosol containers.

The individual trends for the various packaging material groups and their current usage patterns are examined below.

PACKAGING MATERIAL SALES: UK 1972

Pack type	Value £m	% of Total
Tinplate	186	15.5
Plastics	180	15.0
Fibreboard outercases	154	12.8
Cartons	141	11.7
Glass	98	8.2
Paper bags/sacks	80	6.7
Steel drums	42	3.5
Aerosols	42	3.5
Aluminium foil	40	3.3
Cellulose film	35	2.9
Wooden	29	2.4
Jute sacks	26	2.2
Miscellaneous	147	12.3
TOTALS £1200		100.0%

Figure 9.1 Packaging material sales UK — 1972

Paper and board

A study of Figure 9.1 shows that the containers and materials included in this category account for approximately one third of the value of the total packaging market. This contrasts with the situation in 1969 when the comparative figure was almost 50 per cent. During the last few years this sector has been under heavy competition from alternative materials, particularly from plastics. The major development efforts in this area have, therefore, been aimed at keeping down or reducing costs to compete effectively with alternative materials. The other problem here is the reliance of the industry on imported materials. At present, imported materials account for over one third of the UK paper and board consumption.

Fibreboard outercases

These represent the most important part of the paper and board industry, in both tonnage and value terms. In tonnage terms the consumption involves one and a half million tons a year in the UK, equivalent to well over a third of the total paper and board usage. Corrugated board, on the basis of its superior performance to weight ratio, has almost completely replaced solid board for packaging applications. The output of solid board is now estimated at only 5 per cent of the total. A similar pattern is emerging in the USA and Europe. Most of the imported material used in this application consists of kraft lined paper or board. Recently the development of liner boards based on recovered waste paper, known as Test/jute liners, has been successful in reducing the dependence on imported kraft materials.

The replacement of corrugated outercases by shrinkwrapped units, with or without a paperboard tray, has significantly reduced the volume of outercases used for the packaging of consumer products. This is particularly true for the UK where there are significant cost savings to be achieved and where shipping distances are relatively short. This method has been less successful in the USA and Canada where the same cost incentives are not available because the relative prices of corrugated cases are lower and cross continental shipping distances are much longer.

Corrugated outercase designers however, have reacted strongly to the challenge offered by shrinkwrapping. In addition to developing boards with improved performance at lower weights they have also developed improved systems. One example of the latter is the wrap-around case system which takes flat corrugated blanks and automatically forms the outercase around the product to be packed. At the same time the side and end flaps are glued. This method reduces the area of board used for an outercase by up to 20 per cent compared to a regular slotted container and also offers a more efficient packing method. This type of system can easily be adopted to produce an 'easy-open' outer-case which can also serve as a display container in store. Many corrugated outercase suppliers now offer complete packaging systems of this type.

Markets lost to shrinkwrapping have, however, more than been off-set by new applications. This is particularly so in the field of large items such as domestic appliances and industrial products.

The replacement of wooden and metal containers by specially designed corrugated units has resulted in an overall growth rate of 3 to 4 per cent for this sector. Often, corrugated board is used in conjunction with plastic cushioning materials, or with wood, for these applications. Development work is also under way on the production of stronger boards to further improve market penetration in this area. Recent reports from the USA tell of the introduction of quadruple wall containers designed for the pallet load shipment of large engineering items. Studies on the improvement of the chemical properties of corrugated board by the use of plastics, waxes and resins are also under way.

Paperboard cartons

In 1968 a growth rate of 12 per cent was achieved, but this has now been completely arrested. The cessation of growth is attributed to the use of flexible pouches both as an economy measure, (for example, frozen vegetables) and also to improve product visibility. The importance of supermarket selling in this area has led to the upgrading of the board types used, resulting in an overall improvement in finished pack appearance.

This is particularly relevant as nearly 50 per cent of the output is taken by the food industry, and of the non-food areas; cigarettes and tobacco, soaps and detergents, pharmaceuticals and toiletries, the main usage areas, all place major importance on pack appearance. Figure 9.2

A FOOD	'000 tons	% Total
Chocolate/confectionery	59	9.9
Cereals	42	7.0
Flour confectionery	32	5.4
Dried foods/mixes	28	4.7
Frozen foods	24	4.0
Biscuits	14	2.4
Tea	14	2.4
Alcoholic/soft drinks	14	2.4
Ice cream	8	1.3
Margarine fats	7	1.2
Sugar	6	1.0
Milk	6	1.0
Other foods	20	3.3
Total Foods	274	46.0
B NON-FOOD	**'000 tons**	**% Total**
Cigarettes, cigars, tobacco	63	10.6
Soap/detergent powders	52	8.7
Pharmaceuticals	26	4.3
Cosmetics/toiletries	25	4.2
Paper products	22	3.7
Light engineering	18	3.0
Toys and games	18	3.0
Electrical goods	15	2.5
Clothing textiles	13	2.2
Footwear	10	1.7
Other non-food	60	10.1
Total Non-Food	322	54.0
Total	596	100%

Source: *Business Intelligence Services Carton Survey 1970.*

Figure 9.2   Folding carton usage: UK 1970

gives a breakdown of usage of paperboard cartons broken-down by product groups. A further breakdown between food and non-food is also illustrated. Currently duplex type boards and white lined chip boards each hold about a 35 per cent share of the total carton board usage, with solid white board accounting for 5 to 10 per cent and the balance being made up by unlined chip.

As with corrugated outercases, cartoning machine systems are in full use; these handle flat blanks instead of side seam glued cartons. Such methods result in economies in both carton and packing costs. Typical examples of systems of this nature are Kliklok, Sprinter and Diotite. Machinery for these systems is usually supplied by the carton supplier to the user on a rental basis. Rapid progress has been made in the production area with sealing methods. The use of hot melt adhesives for conventional cartons and of hot air sealing for wax or polyethylene coated materials has significantly increased overall line efficiencies.

Developments in composite cartons of plastic and board are well advanced. Cartons embodying moisture barriers using either wax or polyethylene laminations are used for moisture sensitive products such as certain powdered detergents. In some applications the protective medium is applied to the surface of the carton to provide the added bonus of improved appearance, examples of this are polyethylene coated or foil laminated cartons. Carton systems for the packaging of liquids are also in commercial use (for example, milk cartons) and further developments are expected.

Bag-in-box type applications are challenging metal and fibreboard cans, for a wide range of products, particularly in the food area. In-plant conversion exists to a small extent, Boots and Cadburys being two UK examples, but it is not expected to grow significantly.

Paper sacks and balers

The packaging market for multiwall paper sacks is being continually eroded by plastic competition. Large usage areas such as fertilisers and plastic granules have almost entirely changed over to plastic sacks. This trend will continue as the remaining problems with the stacking and handling of plastic sacks are overcome and as the price differential becomes more favourable. Non-packaging uses, such as waste disposal sacks, have taken up most of the capacity caused by losses in the packaging area but this use is also under attack by polyethylene.

Baler bags, which usually consist of two paper plies formed into a bag with a rectangular base have been successfully promoted as cheaper alternatives to outercases for the packaging of products such as bags of coffee and sugar. Machinery is available which produces wrap-around baler bags from paper roll stock for this application. The success of shrinkwrapping has been the main factor in restricting further developments in this area.

Labels/wrappers

Paper labelling still remains one of the cheapest methods of decorating metal, plastic or glass containers and is still widely used. The total volume is, however, being eroded as alternative decorating techniques improve both in appearance and cost.

Food wrappings are another area where competition from aluminium foils, plastics, and cellulose films is continually eroding the total usage of paper. The recent change to the marketing of margarine products in thermoformed PVC containers has also had a significant effect on the volume of glassine and parchment paper used. Future competition from materials such as high–density polyethylene tissue paper is likely to stultify further growth in this section.

Fibre drums/tubes

Fibreboard drums are used mainly for the packaging of powders, pastes and solids, particularly in the chemical industry. They are, however, only suitable for non-toxic products in this area. Fibreboard tubes have a significant usage in the consumer products field especially for food and confectionery. Plastics are also making inroads into these markets, the one for household scouring powders having been almost taken over by thin walled polyethylene containers.

Metal containers

This is the second largest sector of the UK packaging market, accounting in value terms, for a quarter of the total. Tinplate cans are by far the largest container group in this category. In the USA aluminium cans now have a significant share of the metal container market.

Tinplate

Tinplate containers have been subjected to intensive competition in recent years from plastics, aluminium, composite and glass alternatives. Success in the marketing of tinplate containers for the beer and soft drinks market has prevented a major decline in usage. Cost factors are the main stimuli leading to a search for alternative containers.

Tinplate containers maintained a remarkable price stability level from the mid 1960's until 1969, due to developments aimed at keeping manufacturing costs down. Among these were reductions in both tin coating and steel plate weights, together with improved processing methods which gave higher strength to weight ratios. These developments culminated in the introduction of tin-free steel, containers made from steel plate with a protective outer coating of chromium oxide. This method has an added appearance bonus, in that the side seams are cemented or welded, instead of soldered, and hence the can may be decorated all around with the exception of a narrow band.

Since the middle of 1969, however, raw material costs, and hence can prices, have risen rapidly and tinplate containers have been replaced by alternative materials for many applications. One example is the replacement of one gallon (five litre) tinplate containers for the packaging of liquids such as detergents, motor oils and fruit squashes, by blow-moulded polyethylene containers.

Competition from aluminium containers comes in the form of two-piece drawn and ironed cans (see the section below headed 'Aluminium' for details). Processes have now been developed for the production of steel cans by this method. It has the advantage that the metal distribution can be controlled so that the bottom of the can is thicker than the walls, resulting in a strong container from a relatively low metal weight. Complete decoration around the can is also possible in the absence of side seams or welds.

Future tinplate container developments will include improvements to the convenience features. Aluminium ends with 'easy-open' tabs are already widely used for beverage cans. This approach will be extended to ends which tear off completely in one piece. These are already in use in North America for the packaging of nuts and other snack foods. Continued erosion of the overall tinplate market is anticipated as plastics, with improved barrier properties, become competitive in price.

Steel

Drums and pails account for the main usage of steel in packaging. These can be divided into two main groups:

1 Large drums eg 45 gallons (210 litre) and
2 Small drums and pails usually less than 25 litres in capacity.

Large drums The chemical and petroleum industries account for over 75 per cent of the total drum output in this size range in the UK. The drum industry has been affected by the current depression in these two industries and as a result the total sector output has been static. New applications in the areas of food processing and brewing have helped to offset the losses from traditional outlets. As with paper and board, steel has been subjected to heavy competition from alternative materials and systems.

In the case of the large drum, competition from plastics has so far been relatively unsuccessful. One reason for this is price, unlined steel drums (45 gallons/210 litre) cost around £3.50 compared to £5.50 for a blow moulded polyethylene equivalent. Even drums with two coats of a lacquer lining are significantly cheaper, costing around £4.25 depending on the exact coating used. Another factor to be considered when evaluating plastic drums is the efficient reconditioning service which operates in the UK for metal drums, top quality secondhand drums being available at around £1.75 each. Plastic drums have, however, been successful in Germany and it is predicted that with current price trends for steel and plastic, they will become competitive in price in the UK by 1975. Other competition arises in the form of alternative transport systems such as semi-bulk and bulk shipment, particularly in the chemical industry.

The introduction of a new method of drum seaming has been the major development in this area in the last year. This was pioneered by Van Leer and consists of producing a spiral or round seam instead of the conventional sharp edge. This has resulted in an overall increase in container strength and will ultimately lead to the specification of lower weight containers. Originally this development work was initiated to bring the drum performance up to that required by international transport regulations for the shipping of hazardous products. Van Leer have also introduced a lightweight one-trip drum which conforms with these regulations (RID).

Small drums and pails   Blow-moulded and injection-moulded plastic containers have currently taken over large usage sectors from steel containers. The drum companies in the UK are all represented in this section of the market.

For steel drums which are still used for the packaging of products such as aromatic chemicals, lightweighting has been carried out in an attempt to offset rapidly rising steel prices.

Aluminium

The usage of aluminium for packaging is still relatively low, the main area being that of foil, particularly for wrapping or as part of a flexible laminate. The main potential market is as a replacement for tinplate containers, particularly for beverage cans. In this market, aluminium containers are still at a price disadvantage in the UK. Even though tinplate container prices have risen fast in the UK lately, the can costs are still relatively low compared to those in other countries. In the USA the price differential between tinplate and aluminium is lower than the UK, and the aluminium suppliers have cut prices in order to obtain an 8 per cent share of the market.

In the long term, aluminium containers produced by the drawn and ironed process or by extrusion are expected to become competitive with tinplate, particularly for small containers. The drawn and ironed process consists of taking the aluminium alloy in sheet form and feeding it into a press where it is shaped into cups. The cups are then drawn out and ironed to reduce the diameter and achieve the necessary height. The cans are then trimmed, cleaned, printed and flanged for shipment. Aluminium cans of this sort are stackable which is a plus point in the market-place and also a help during storage.

Metal tubes and moulded trays are the other two main packaging uses for aluminium. The main markets for tubes are toothpastes, pharmaceuticals and cosmetics. Plastic tubes are, however, making progress particularly in the latter area. Moulded aluminium trays are used in the food industry for convenience foods in both consumer and institutional outlets.

Aluminium containers also have a share of the top end of the aerosol market. One major plus point for aluminium in this area is its high corrosion resistance.

Aerosols

In 1972 world production of aerosols is expected to exceed 5,000 million. This is the fastest growing container group in the UK at approximately 20 per cent per year. This growth has been achieved despite adverse economic conditions and steadily rising container costs. Tinplate is the main container type although glass, aluminium and plastic aerosols are all marketed. The main uses are hairsprays, household products and personal products. In the USA hairsprays have been overtaken as the major category by both household and personal products.

Figure 9.3 shows a table illustrating the growth of product categories for Western Europe from 1964-1970. This shows that the main growth areas are personal products, particularly deodorants and anti-perspirants, and household products. The success of slow release plastic strips for insecticides has caused a slowdown of the growth of aerosols for this application. Food aerosols account for only a small part of the total. Even in the USA, where much development work has been carried out,

AEROSOL CONTAINERS: WESTERN EUROPE

GROWTH OF MAJOR PRODUCT GROUPS 1964-1970

	1964 Million Units	1966 Million Units	1968 Million Units	1970 Million Units
Insecticides	94	108	135	140
Hairsprays	199	314	375	510
Personal	63	115	210	355
Household	111	128	170	255
Paint	13	20	35	40
Other	50	65	95	125
Totals	530	750	1020	1425

Source: *Aerosol Figurama – Metal Box Company Ltd. 1972.*

Figure 9.3 Aerosol containers – Western Europe – Growth of major product groups – 1964-1970

they comprise less than 5 per cent of the total aerosol market.

Cost is not such an important factor in development programmes as it is for other container groups, although continual progress is being made with cheaper valve components, container specifications, propellants and product formulations. The success of water based formulations has significantly reduced costs for many applications. The main development effort is concentrated on improved performance and new areas of application.

Several types of co-dispensing aerosols are now being marketed. These keep potentially reactive components separate until they are emitted, one example being hot shave foams. However, formulation development work on systems such as microencapsulation may result in this type of formulation being packed in conventional aerosol containers. It is predicted that the general usage pattern will follow that already established in the USA.

Plastics

Packaging is the largest single market for plastic materials; accounting for approximately 25 per cent of the total output. The growth of plastics in packaging has been rapid in the last few years, averaging about 20 per cent per year. The UK and European markets have developed more rapidly than the American. The 15 per cent market value accounted for by plastics in the UK (see Figure 9.1), compares with a figure of around 6 per cent in the USA. Prices of plastics started to rise in 1966-67 after a long record of reductions. However, future rises are not expected to be high due to current overcapacity in the industry.

Low-density polyethylene has always been the leading polymer used in the packaging field and it still maintains a 40 per cent share, despite competition from high-density polyethylene, polystyrene, polyvinyl chloride and polypropylene. Figure 9.4 shows a breakdown of the UK consumption of plastics for packaging in 1972. The main usage area for low-density polyethylene is for shrinkwrapping and this continues to expand. The usage of high-density polyethylene and polyvinyl chloride is growing rapidly for blow moulding applications. Polypropylene usage is expanding on four fronts; injection-moulded crates and trays, film, blow-moulded bottles and twines and string. The rapid development of

CONSUMPTION OF PLASTICS FOR PACKAGING: 1972

Material	Volume ('000 tons)
Low density polyethylene (LDPE)	198.0
High density polyethylene (HDPE)	71.0
Polystyrene (PS)	69.0
Poly vinyl chloride (PVC)	54.5
Polypropylene (PP)	48.5
Polyvinylidene chloride (PVDC)	6.0
Thermosets	7.0
Miscellaneous	2.0
Total	456.0

Figure 9.4   UK consumption of plastics for packaging – 1972

techniques in the areas of thermoforming and foamed cushioning applications have helped polystyrene usage to increase by over 50 per cent since 1968.

In following paragraphs the usage trends and future developments will be considered under production method headings.

Blow moulding

Polyethylene is the traditional material used for this process in the UK. In the last few years polyvinyl chloride has established a significant share of the market for applications such as fruit squashes, disinfectants, shampoos, liquid detergents and recently spirit miniatures. The breakthrough of this material into the food market came with the development of octyl-tin stabilisers. There has also been a trend from low density to high-density polyethylene for blow-moulded containers, for example for liquid detergents.

One large potential market for plastic containers is the packaging of milk. Currently test markets are in progress in two areas of the UK with lightweight polyethylene containers. In the long term the usage of polypropylene and other higher polymers is expected to increase. Work is being carried out with the use of modified acrylics in those areas which require superior barrier properties as well as good impact strength and clarity. Test markets with this type of material for carbonated soft drinks are underway in the USA. However, the costs are still significantly higher than for the equivalent glass container.

On the machinery side, the use of injection blow moulding for both small containers and wide-mouthed jars is growing fast. This method finds particular application for packaging of cosmetic products where close tolerance moulding is required. Development of a wide range of equipment for the in-plant captive type of operations is also well underway. This applies to machinery suitable for low volumes as well as the larger runs.

Several methods of in-plant moulding involving a form/fill/seal type of operation are also on the market. In these processes the bottles are blown, filled with product and sealed, while remaining in the mould. This method is particularly applicable for liquids which require aseptic filling.

At the other end of the container size scale, rotational moulding is competing with conventional blow moulding for the production of large items such as barrels and tanks. This process consists of introducing the plastic in powder form into the mould and applying heat, at the same time rotating the mould. The powder melts and is deposited on the mould surface. After cooling, the finished article is removed from the mould. It has cost advantages in terms of both initial mould cost and unit cost when compared to blow moulding and is particularly suitable for short production runs of large items.

Injection moulding

There are three main usage areas for injection moulded packaging materials; crates and trays, thin-walled food containers and closures. The change-over to plastic crates, made from either polypropylene or high density polyethylene, is almost complete in the milk market and is well underway in the brewery trade. Polypropylene now has the majority share of this application. Systems for the handling of foodstuffs such as fresh produce, bread and bakery products, involving the use of injection moulded trays are also increasing fast.

Injection moulded polystyrene containers form the main pack type for convenience dessert products such as yoghurts and mousse, and are also used for products such as cream, cheese spread and butter portions. Recently a process has been patented in the UK by Airfix Industries for the production of thin-walled injection-moulded shells which combine with a wrap-around label to give a low priced thin-walled container suitable for products such as yoghurt. Injection-moulded closures are now in wide use for many product areas. They have replaced both thermoset plastics and metals for closures on performance and cost grounds.

A novel injection-moulding process known as the structural foam process has recently been introduced. This consists of injecting a mixture of the plastic plus a chemical gassing agent into a heated mould. The chemical breaks down under the heat and evolves a gas (usually nitrogen) which causes the plastic to foam. The pressure exerted during this type of process is low compared to conventional injection moulding and hence a cheaper mould can be used. The foamed structure produced is more rigid than an unfoamed one of equal thickness and also contains less material.

Extruded films

These account for almost 45 per cent of the total plastics packaging market and are the main outlets for low-density polyethylene. The main areas of application are shrinkwrapping, overwrapping, plastic sacks, bags and flexible laminates. These are described below with the exception of flexible laminates which are discussed in a later section.

Shrinkwrapping   This was the packaging innovation of the 1960's. The method, consisting of replacing corrugated outercases by a heat shrink overwrap of film, is particularly suited for the needs of supermarket and cash and carry operations. Equipment development has now caught up with film technology and complete systems are now offered over a wide range of speeds.

The market is broken down between low density polyethylene and polyvinyl chloride on a 3:1 basis. The more expensive polyvinyl chloride is preferred to polyethylene for some applications because of its superior appearance. Shrinkwrapping of palletised loads with heavy duty polyethylene film is being applied to an ever increasing range of products. Again a wide range of manual, semi and fully automatic equipment is available.

Overwrapping   Regenerated cellulose film, commonly known as 'Cellophane' (British Cellophane trade mark in the UK) is the traditional material used for the overwrapping of products such as biscuits, cigarettes and chocolates. Orientated polypropylene film has made significant inroads into this market in the last few years. Initially progress for this film as an overwrapping material was slow because of technical problems. These involved, firstly, variable film quality and secondly, problems with machine running, as more controlled conditions of temperature and pressure were required compared to cellulose film. These have now been solved by improved processing, by modifying current equipment or by the purchase of new wrapping machines. Currently, cellulose film prices are rising more steeply than those of polypropylene, and a complete market domination for the latter is forecast. As with cellulose film, a wide range of coated and uncoated grades of polypropylene film are available.

The usage of thin high-density polyethylene films as a replacement for food wrappings is also spreading fast. These materials have the advantages of high wet strength, grease resistance and dead fold properties, all at an economical cost.

Sacks   The story of plastics replacing traditional materials repeats itself in this area. Low density polyethylene sacks have now taken over major markets such as those for fertilisers and plastics raw materials from multi-ply paper sacks. Both open mouthed and valved sacks are used for these applications. The valved sacks have advantages for handling and palletising operations.

Often, palletised loads of plastic sacks are shrink film overwrapped for extra stability and protection, particularly for export shipments. For heavy duty sacks, the traditional materials such as jute have almost completely been replaced by woven polypropylene and high-density polyethylene. The latter comes in the form of Van Leers 'Valeron' and consists of two sheets of orientated high-density polyethylene film, which are cross laminated for extra strength. Both the back seam and end seals are adhesive bonded. This sack has significant performance advantages to alternatives in the distribution system.

Bags   Low density polyethylene is the pacesetter in this sector and the usage is still growing rapidly. This trend is expected to continue as paper rises in price. Some of the main products packed in this way are food products such as meat and vegetables and confectionery. Film bags are also gaining in popularity as take-home units in supermarkets.

Polypropylene film bags have established significant market segments in the food industry for convenience snack foods (crisps etc) and in department stores for the packing of clothing. In both these applications the attractive appearance of the polypropylene film is used to boost sales.

Thermoforming

Thermoforming consists of the production of a packaging item or container from a sheet of plastic. The sheet is heated and is shaped around a mould either by drawing a vacuum under the mould or applying air pressure on top. Much of its current popularity is due to its inherent suitability for economical in-plant operations. Significant savings are achieved by the user company both in terms of unit pack cost, and in storage space.

The main usage areas for thermoformed items in packaging are blister packs, trays and thin-walled inserts (for example, for cakes and chocolates). A wide range of plastics materials including polystyrene, polyvinyl chloride and high-density polyethylene are used for thermoforming. Co-extruded films can also be thermoformed.

In the packaging area, the main expansion of thermoforming is for the production of unit containers. One of the major successes in the food industry recently has been the introduction of soft margarine packed in thermoformed polyvinyl chloride tubs. Further developments in this area involve the combination of thermoforming and spin welding techniques to produce unit containers. Two halves of a container are produced on a thermoforming machine, delivered to the user, who then spin welds them to provide the finished container ready for packing. Storage space of the two halves of the container is much lower than for completed containers. Complete in-line operations utilising this process are now being offered in the USA. The method is also suitable for the new acrylic materials. One of the test markets of soft drinks packed in acrylics in the USA is produced in this manner.

Other machinery developments concern the use of cold forming techniques. These have advantages of speed, lower equipment cost and reliability all due to elimination of the heating process.

Glass containers

Glass containers, the fourth largest material group (see Figure 9.1), have maintained their share of the total packaging market in recent years, growing from £61.2 million in 1966 to £98 million in 1972. This has been achieved in the face of competition from metal, plastic and flexible alternatives. There are two main reasons for this growth rate:

1 The success of cost reduction programmes in the glass container industry
2 The development of key markets for one trip containers.

Cost reductions have been achieved by improved processing methods and by significant reductions in container weights. The latter was made possible by the development of methods of applying chemical coatings to the glass surface. The coatings improve the impact strength of the glass container both on the filling line and in use, and allow lower glass weights to be used. The coatings are applied either after the container leaves the mould or after annealing. They are described as hot or cold end coatings depending on which end of the annealing lehr they are applied. A wide range of coatings are used depending on the containers performance requirements, some examples are sulphur, soap, metallic oxides, waxes and resins. A typical combination is the application of tin or titanium oxide at the hot end and a polyethylene dispersion at the cold end. Recent developments include the use of an organic metallic compound as a hot end coating, which undergoes a diffusion reaction, resulting in a penetration of the metal into the glass, giving a significant increase in strength. The success of such lightweighting techniques has meant that price increases have been kept to a minimum and glass containers have remained competitive with alternative materials.

A rapid growth in the use of one-trip containers for the carbonated soft drink and beer markets is the second major factor accounting for the steady growth rate of glass container usage. One of the reasons for the rise in popularity of the one-trip container is the reluctance of supermarkets to handle returnable containers. The easy-open metal can is challenging strongly for this portion of the non-returnable market and there are signs of a change from glass to metal, particularly in the USA. One advantage of the metal can is the high filling speeds which are achieved. In the USA the Glass Containers Manufacturers Institute has recently sponsored a project to develop a filling line for non-returnable carbonated soft drink glass containers to run at 2000 per minute.

The chemical inertness of glass is one of its main advantages as a packaging material. This is a definite plus for products requiring a long shelf life. It is also an excellent barrier material. The other important feature of glass from a marketing viewpoint is its transparency. In Europe, for example, many food products which are canned in the UK, are packed in glass for added sales appeal. One disadvantage of glass is that, in its normal flint form, it does not give protection against light. This is overcome for certain products by the use of amber glass.

The properties of glass have resulted in its usage in main markets such as food, dairy products, chemicals, pharmaceuticals and alcoholic drinks. Figure 9.5 shows a breakdown of glass usage by product group for the UK from 1968-70. The increase in the major usage area (food), in 1970 was accounted for mainly by rapid growth in the sales of glass packed baby food and instant coffee. In both these sectors the change was from metal cans to glass containers. The increases in the soft drinks and beer/cider markets are accounted for by the increased usage of one-trip containers. The falling usage pattern for glass containers in the

USAGE OF GLASS CONTAINERS : UK 1968-70

Product Group	1968 Million Units	1969 Million Units	1970 Million Units
Food	1787	1740	1878
Wines and spirits	902	920	988
Soft drinks	754	908	991
Chemicals/pharmaceuticals	700	668	715
Dairy products	446	486	503
Beer/cider	241	303	323
Toiletry/cosmetics	402	363	325
All other	420	477	520
Totals	5652	5865	6243

Source: *Glass Manufacturers' Federation*

Figure 9.5   Usage of glass containers UK — 1968-1970

toiletry and cosmetic sectors reflects the effect of plastics competition.

The major new development in glass packaging is that recently completed by Owen-Illinois in the USA of a composite glass/plastic beverage container. This has now reached the market after an extensive testing programme. The process consists of extruding molten glass as a ribbon and then blowing it into a pear shaped container, which has a wide mouth. This is then welded to a plastic base. The finished container is stable, easy to fill (and drink from) and despite its thin walls, very strong. It is also easy to crush for disposal.

Future developments will include further attempts at lightweighting, particularly to meet the challenge of plastics. Methods such as chemical tempering, where improved strength is obtained by an ion exchange process at the glass surface are actively being developed. Production developments in further automating and speeding up the blowing process and the quality monitoring systems are expected to progress. A long term target of many suppliers is the production of predecorated containers which will be competitive in price with labelled packs.

Flexibles

Packaging materials, supplied in roll form, which are used for the in-plant production of packaging items are covered by this heading. A wide range of materials such as paper, aluminium foil, plastic, cellulose and rubber films, together with many laminate combinations are employed. Some of these packaging materials have already been discussed in the previous sections of this chapter.

There are no statistics available on the total usage of flexible material used for packaging in the UK. In the USA flexible packaging sales account for approximately £600 million per year, equivalent to 7½ per cent of the total market. Available UK statistics cover the use of flexible laminates in packaging and are published quarterly in the Department of Trade and Industry *Business Monitor*. The value for 1972 is estimated at £15 million equivalent to 1¼ per cent of the total market (see Figure 9.1). The main laminate components used are paper, aluminium foil, polyethylene and cellulose films. Usage of aluminium foil, is high mainly because of its good barrier properties. Polyethylene finds a wide application as the sealing medium on the inner ply of many laminates.

There are two main types of machinery used to form, fill and seal flexible materials, namely vertical and horizontal units, depending whether the web feed is in the vertical or horizontal plane. They produce overwraps, pouches (with or without gussets), bags and even more elaborate shapes such as tetrahedrons.

One result of development effort has been the introduction of machines suitable for the gas packing of oxygen sensitive products. These consist of either replacing the air in the pouch with an inert gas (usually nitrogen) flush or by drawing a vacuum and then injecting the inert gas. The former method, which is capable of producing pouches with oxygen levels of less than 1 per cent, is preferred on grounds of higher output. Food products now packed in this way include coffee, snack foods, peanuts and dried yeast. This packaging system together with a coated cellulose film/polyethylene laminate gives a finished pack with a shelf life of up to three months. Machines are also available for the pouch packing of liquids such as milk, fruit drinks, and oil. These often involve aseptic filling conditions in which the roll stock is treated with hydrogen peroxide prior to forming, and the filling is carried out under hot sterile conditions.

In addition to the widely used laminate components mentioned previously higher cost materials such as polyvinylidene chloride, the polyamides and polyester are finding increasing usage for specialist applications. Single material films such as polyurethane, polybutylene and thermoplastic rubber are being screened for potential usage areas. Recently, a UK test market of a pouch-packed beer has commenced, using a construction of biaxially orientated polyethylene teraphthalate film coated with polyvinylidene chloride.

Development work is also proceeding on co-extrusion techniques as methods of producing cheaper, more flexible alternatives to laminates. A wide range of plastics materials are being evaluated particularly in the USA and Japan. For the packaging of oil for example, co-extrusions of polyethylene/polyvinylidene chloride/polyethylene are being evaluated. Development of co-extrusions of ionomer and polyethylene for medical applications are also well under way. This type of strong film has the advantage that it is permeable to sterilising gases such as ethylene oxide.

The disadvantages of co-extrusions are:

1 They must be surface printed, compared to sandwich printing for conventional laminates and

2 Any scrap produced during processing cannot be recycled.

The second disadvantage is overcome by a new method of composite plastic film production known as alloying. In this system two plastic films are combined into a homogenous mass, which is capable of recycling.

The main development effort in the area of flexibles is directed at producing a pouch type alternative to the metal can. Test markets are under way in both the UK and the USA with flexible pouch packs of food which have been retorted after filling. The materials used are laminates of polyester/aluminium foil and nylon or polypropylene. One of the main problems with the development of a retortable flexible can is the use of aluminium foil as a barrier material. While its barrier properties are excellent it is brittle and tends to develop fine cracks when handled excessively. Alternative materials such as polyvinylidene chloride do not provide adequate barrier properties for the flexible can. Polyacrylonitrile is a potential material for this application, providing methods can be found to produce it efficiently.

The examples discussed in this chapter illustrate the rapid changes occurring with packaging materials and machinery. This chapter has covered the five major material groups but there are several other material categories in which similar changes are evident. The story in all material groups is of a challenge between the traditional materials and the newer materials such as plastics. The traditional materials are reacting to this challenge by implementing cost reductions and finding new areas of application. The plastics materials are continually searching for new applications and also rapidly developing new processes and new polymers. The result of this competitive activity is more efficient packaging systems in many areas.

10

Packaging as
a Marketing Tool

Marketing can be defined as the overall strategy or complex which moves goods from the source of production into the hands of the consumer. Within this definition several functions can be identified; these are exchange (buying and selling), supply (transport and storage), standardization and grading, financing, risk taking, and the provision of market information. Packaging is deeply involved in many of the above functions and should be taken into account at an early stage in any marketing plans.

The importance of the pack in relation to the product is generally accepted insofar as luxury or semi-luxury items are concerned. In the cosmetics industry, for example, the pack plays a vital part in promoting sales, and a great deal of time and money is expended on pack design. A similar situation exists in the expensive end of the chocolate trade, especially for gift packs designed for seasonal promotions at Christmas and Easter.

The pack should be recognised as an important marketing tool in many other fields, however, especially in the light of the increasing emphasis being placed on self-service selling, advertising and sales promotional schemes.

The importance of packaging as a marketing tool will be seen more clearly if its various interactions with other marketing functions are analysed in more detail. The most important of these will be discussed under the following headings:

1 Retail marketing trends
2 Design and company/brand image
3 Sales promotion
4 Advertising
5 Distribution

Retail marketing trends

Early trends

The growth of packaging has had a tremendous effect on retailing methods. Packaging, in its turn, has itself been affected by changes in retail marketing. The first major change was, of course, from a state of minimum packaging (and even that carried out in-store) to one where a majority of items were packed and, of these, most were packed by the manufacturer. There was a gradual transition during this period (say up to the beginning of World War II) from packaging which was meant only as a container, to packaging with a definite emphasis on sales appeal.

Packaging had two main effects on retail marketing during this period. It enabled manufacturers to sell more and more goods under brand names and it changed the character of the retail shops. The interiors of shops became less cluttered, while allowing the display of a greater number of products, and window display was revolutionised. The window of the normal chemist's shop, for instance, used to contain not goods, but glass vessels filled with coloured water, with perhaps a pestle and mortar or a pair of apothecary's scales. The advent of packaged goods enabled the contents of a shop to be displayed in the window as an inducement to fresh customers.

In the main, these changes were brought about using old established packaging materials such as glass, tinplate and paper/board but there was one new material which contributed a great deal to sales appeal, and that was cellulose film. There is little doubt that its combination of clarity, colour (where necessary) and protection against moisture vapour and liquid water did a great deal to speed up the packaging revolution.

Self-service and supermarkets

Since the end of World War II, the biggest change in retail marketing has been the growth of self-service. Self-service on any large scale is completely dependent on packaging although it is also true to say that the growth of self-service stores and supermarkets has had a correspondingly great impact on packaging developments. Having once established the concept of supermarkets it was natural to bring under their wing the marketing of products not normally packaged. One large area affected was that of fresh fruit and vegetables.

The desire to round out the product range with fresh produce led not only to developments in the packaging of such commodities but also led to changes in distribution. Certain chain stores and supermarkets have their own central pre-packaging facilities and so buy in bulk either from local farmers or from the usual fruit and vegetable market. Other supermarkets and self-service stores buy from packing stations who are themselves supplied with produce from a number of farmers in their locality.

Supermarkets also had a direct effect on the packaging of confectionery. At one time most sugar confectionery was sold unpackaged but the practice of placing confectionery near check-out points of supermarkets for impulse sales soon led to the packaging of toffees, boiled sweets and other items in cartons or film sachets, in order to obtain the benefit of increased sales.

One important influence of supermarket selling on packaging is on the graphics of the package. In the self-service environment the package acts as the product's only salesman and quite small differences in shape or surface decoration may mean the difference between sale and no sale. The question is not just the simple one of designing a package which will stand out from all competitive packs.

There is often a direct conflict between designing a package which will attract the maximum amount of attention and one which will project the correct image. If the desired image is a brash, 'with it' one aimed at the 'pop' generation, all well and good, but this is not always the case. The use of a completely different shape may be one answer but this may bring problems in stacking on the supermarket shelves. Selling space is an extremely valuable commodity in a self-service store and the store manager is not anxious to stock products which make inefficient use of his valuable space.

It must also be remembered that many stores market a range of commodities under their own private label. Anyone else's product has to prove that it is capable of earning the store owner a good profit before he is prepared to give it space on his shelves.

Supermarket selling has another influence on package decoration. Because of the emphasis it places on the package acting as salesman it is essential that brand recognition is possible from as many different viewpoints as possible. In other words if the staff who fill the shelves happen to place your carton or bottle the 'wrong' way round is it still recognisable?

To summarise, the most important factors for a successful package for supermarket selling are as follows:

1. The package must be convenient to stock and display.
2. The package must have an attractive appearance.
3. The package must be capable of preserving the contents during storage and display.
4. The package should not be easily soiled.

Vending machines

Vending machines constitute another selling area which is almost entirely dependent on packaging for its existence. The emphasis here is more on the performance side of the package rather than on sales appeal since the package is usually largely hidden. Maximum utilisation of storage space is even more vital in a vending machine than on a supermarket shelf. Protection of the product is paramount since the customer is likely to be even more annoyed than usual if he receives stale goods, because he is unable to complain on the spot. He is, therefore, far more likely to make a mental note not to buy your product again.

The benefits of vending machines are, of course, the possibilities of much wider sales through a greater geographical coverage and a theoretical twenty four hour opening time.

Cash and carry

Cash and carry outlets serve the smaller shopkeepers and cut down the need for low profit calls by salesmen to small outlets. The recent growth of cash and carry outlets in the UK, for example, has been dramatic, and proceeds were estimated at £52 million per year in 1971 and are expected to reach £100 million by 1975.

Cash and carry chains are also expanding throughout Europe. Attractive outer packaging is necessary to display the product in these outlets, and shrink wrapping has found a ready acceptance as an outer pack because of its good display value.

Another demand that cash and carry operations make on the outer packaging is ease of break-down into smaller units. Orders in this type of outlet usually vary between £10 and £250 which means that some items may be required in small numbers — sometimes even as low as threes and sixes. In the case of shrink wrapping, this means that a pack of, say, two trays — each of twelve jars or cans — shrink wrapped with polythene film, will have to have the film removed in order to sell six containers. The tray, therefore, must be sufficiently rigid to be handled on its own without spilling the contents over the floor.

Design and company/brand image

One of the definitions of packaging given in Chapter One was: 'A package must protect what it sells, and sell what it protects'. Another way of stating the second half of this definition is to say that the package must have 'sales appeal'. Implicit in this is the fact that the package is an important means of creating a favourable brand or company image.

Before considering the creation of a favourable image for a new product, a decision should be made on whether the product is to be sold by brand identity or corporate identity. In most cases the answer will lie somewhere in between these two extremes but examples also exist of successful exploitation at either end of the scale. Leading exponents of the successful marketing of products under brand names are the large detergent manufacturers, such as Unilever, and Procter and Gamble. Few housewives who buy their favourite brand of 'whiter than white' could tell you the name of the manufacturer. At the other extreme there are companies such as Schweppes and Heinz who sell all their products under the company name and do it very successfully. One thing is not in doubt. Whether brand or corporate image is being promoted the package now plays a major part. The package may create a favourable image in two ways:

1 By presenting a pleasant visual image via intrinsic shape or surface decoration.
2 By correct and efficient functioning.

Design/decoration

The creation of an image by the package is a complex matter and is easily affected by what might appear to be minor changes. Even changes in colour can affect the product image in some instances. In the field of baby care items, for example, there still remains something of the tradition of 'pink for a girl and blue for a boy', and the package often reflects this. In food packaging, too, colour can have an effect.

A trial conducted with coffee packed into tins having one or other of three differently coloured labels, produced some very interesting results. The labels were coloured either brown, red or yellow and were otherwise plain, with no copy or other distinguishing marks. Coffee from a uniform batch was placed in the tins and tasting tests were carried out among housewives who were allowed to know from which tins the samples were taken. In almost every case the housewives chose as the strongest brew, the coffee from the tin with the brown label. Similarly, that from the tin with the yellow label was usually reckoned to be the weakest.

Container shape is also a factor in creating a product image although account has to be taken of what may be conflicting requirements by the consumer and the retailer. To the consumer a tall, slender bottle with graceful curves may suggest the ideal image for a beauty care product but the retailer may dislike it intensely because of the difficulty of stacking it on his shelves. This difficulty may often be overcome by putting the bottle in a carton and then using surface design to suggest the image of gracefulness and beauty.

An important part of the surface decoration is the logotype. This may simply be the brand or company name, or it may also incorporate a design (such as the sword motif, associated with Reckitt's pharmaceutical products). However successful a product may be, there eventually comes a time when the image needs to be brought up to date. Pack design and logotype design are key factors in any such modernisation. Slight changes in both of these can be sufficient to update the image without loss of identity.

Performance

It would seem to be a truism that package performance can make or mar the product (and indeed the company) image but one still encounters many examples where this truism has been ignored. A beautifully decorated container which nevertheless leaks a highly perfumed product into the housewife's basket is not exactly promoting the company image, particularly if it happens more than once. This aspect is part of the larger subject of 'Packaging and Protection' which was dealt with in more detail in Chapter One.

There are other factors of performance which also affect the product image. One of these is dispensing. In extreme cases, such as the aerosol container, the container is built around the concept of dispensing the product in a certain form, such as an aerosol mist, a wet spray or a foam. However, even in more mundane packaging items the question of dispensing may make or mar the product image.

Finally, there is the question of disposability of the package. This is a matter of increasing importance and one which may well have important repercussions on the brand or company image. This subject will be examined in greater depth in Chapter Twelve; 'Solid Waste Management and Packaging'.

Sales promotion

By promotion is meant some special or non-routine extra effort made to give sales of a product a special boost. A promotion is essentially, then, a short-term sales programme and is sometimes tied to a particular marketing event such as Christmas or Easter, or to a season of the year. Promotions may also be mounted in order to introduce a new product. The eventual target of all these is increased sales, and packaging is often vital to the success of the campaign. Some of the promotional package ideas which can be used are listed below.

1 Money-off pack
2 Bonus size package
3 Coupon pack
4 Pack-on premium
5 Pack-in premium
6 Premium package
7 Self-liquidator

Money-off pack

This has the advantage of being easy to execute and of adding very little to the package cost. In fact, if the 'flash' announcing the deal can be printed using no additional colours then the package cost is unchanged and the only extra cost, overall, is the cost of design and new printing plates. On balance, however, it is often preferable to use an extra colour for the 'flash', usually a bright, contrasting one. The disadvantages are that it can weaken brand loyalty and it is easily and quickly copied by competitors. In spite of this it remains a popular and successful form of promotion.

Bonus size package

Here the price of the standard pack is retained but the size of the pack is increased. The extra packaging costs will vary with the type of package considered. A larger carton, for instance, will add merely a small cost for the extra amount of board used, plus costs for design and printing plates. A larger glass or plastic bottle, on the other hand, will add extra mould costs, the extra cost of a larger bottle and, perhaps, a cost for a new size of closure. Advantages and disadvantages, in general, are the same as for the money-off pack.

Coupon pack

In this promotion, a coupon bearing a stated value, redeemable against the purchase of a product, is packed into a container of the same product (usually) or another product. A variation on this is to make the coupon part of the package; for example, printed on the label of a jar, or printed on a carton. Costs include the cost of a coupon (if separate), printing plate costs if printed on a carton or label, cost of insertion into a package, and redemption costs (either handling by the company at its head office or reimbursement of dealers).

Advantages are that it builds up brand loyalty and it can be used either to introduce a new product or to create demand for a larger size pack. On the whole, though, it is less popular than the straight money-off promotion.

Pack-on premium

This promotion consists in attaching an article having some value or utility on to the standard package. One advantage here is that the basic package need not be changed, while another is the fact that the appeal of the premium offer is enhanced by its visibility. There is, however, the risk of pilferage and the extra cost of the package may be high in relation to that of the premium offer.

One of the factors leading to the extra packaging cost is the fact that a new outer case is necessary. One way of attaching the premium to the product is by shrink wrapping. This ensures maximum visibility and does not add greatly to the cost.

When mounting this type of promotion it is important not to use poor quality premiums as these reflect on the quality of the product.

Pack-in premium

This is similar to the above but the premium is placed inside the standard package. It is not feasible in many cases, such as with bottles of liquids or canned liquid products, and is most effective if the premium offer bears some relation to the product or its use.

A limiting factor in this type of promotion is the compatability of the premium article with the product, particularly if this is a foodstuff. Most moulded plastics are suitable but anything which is printed (such as small booklets, picture cards, etc) should be thoroughly tested. In case of doubt they should be kept out of direct contact with the food either by enclosing in a film bag or by placing between, say, the carton and its inner lining (as with picture cards in cartons of tea).

Premium package

A premium package is one which has some re-use value after the original contents have been used. One simple example is instant coffee packed in glass tumblers (with suitable closures) while another is bath salts packed in an apothecary's jar. The latter example would be a suitable premium to mount at some holiday or gift season. In this type of promotion the package is obviously all important. One advantage is that it can be made to be a constant brand reminder during its period of re-use. On the other hand it is not a promotion that can be mounted quickly and it may incur high investment costs.

Self-liquidator

In this type of promotion the idea is to give the public the opportunity to buy goods at a reduced price. The condition of the offer is that the customer sends in, in addition to the money, a certain number of labels, stoppers, tokens or some other evidence of having bought the package being promoted. The role of the pack in this type of promotion is to carry some distinguishing portion which can be sent with the money by the customer. This should involve no more cost than the cost of design and extra printing plates.

Advertising

The main advertising media are:

1 Commercial television
2 Commercial radio
3 National and local press
4 Trade and technical press
5 Display material

The important point to note about the different media is the way in which they make their particular impact. Television, for example, combines visual and audio impact, whereas radio has audio impact only, and the last three media have visual impact only. Television is the dearest medium but can also, of course, have the greatest effect.

It is also necessary to be quite clear which audience you are aiming to reach. Television, radio and the national or local press are ideal for advertising consumer goods but industrial goods are better advertised in the trade/technical press. It is also possible to mount some industrial advertising in the prestige national dailies, on the grounds that they are widely read by top management.

Successful packaging can be a very useful part of advertising but it is essential that the images projected by the package and by the advertising material should be completely integrated. Confusion can easily occur in the consumer's mind if the package design does not agree with the image built up by advertising.

A particularly important point, here, is to take account of the particular advertising media to be used, when designing the pack. A design which is effective when seen in full colour on the retailer's shelf may have very little impact when presented in the pages of a black and white advertisement or on black and white TV. Similarly with shape. A design which stands out on the supermarket shelf may be almost unrecognisable in a 2-D representation.

Correctly designed, the package should be considered as an advertising medium in its own right. After all, it makes its appearance on the retailer's shelf, in the housewife's basket and often for an appreciable period of time in the home. A designer should take full account of this if maximum value is to be extracted from the package.

To a certain extent the above remarks apply also to industrial packaging. The advertising in trade journals, leaflets at exhibitions, etc, should tie-in with the package. Even if the package is designed to be strictly utilitarian, it is nearly always possible to mark it with some recognisable symbol such as the company logo.

If the package is to be an efficient marketing tool as an adjunct to advertising, it must be instantly recognisable. Sometimes this can be achieved by a distinctive shape of container but in most cases, identification will rely on surface decoration of the carton, bottle, tin, etc. In general, surface decoration will be carried out either by labelling or by printing.

Labelling

Labelling can be the cheapest way of identifying a package although not necessarily so. Decorative labels, for instance, can be quite expensive, or it may be necessary to have labels resistant to spillage of the contents during filling, or to environmental factors such as heat or high humidity. All these requirements will increase the cost of the label. One disadvantage of labels which must be taken into account when making a choice of identification method is that labels can come adrift during transit and storage (particularly under some export conditions) and identification, together with customer goodwill is lost.

Cost of labels depends on a number of factors such as the type of paper, type of printing, numbers off, etc, but in general the four main categories of paper labels can be arranged as follows in ascending order of cost:

1 *Plain paper labels*. These are applied after addition of an adhesive, at the point of application, to the container.
2 *Pre-gummed paper labels*. Here the paper is pre-coated with dextrin or gum arabic and applied after addition of water.
3 *Thermoplastics paper labels*. The paper is pre-coated with a synthetic thermoplastics resin and applied after activation of the resin by heat.
4 *Pressure-sensitive paper labels*. These labels are pre-coated with a permanently tacky adhesive, protected with a separate backing paper, and are applied by the application of pressure.

Printing

Printing of a package is affected by a number of factors, the most important being the package to be printed (its material and shape), the printing process used and the type of ink used. These factors are to some extent interrelated and it is convenient to consider the various printing processes and indicate the packages for which they are most suited and some idea of the inks normally used.

The main printing processes used in packaging are:

Gravure The gravure process consists of revolving an etched metal roller in an ink reservoir. The ink is held in the recesses of the etched design, and the excess ink is scraped off with a steel doctor blade. Reel-fed material (paper or film) is pressed against the etched roll by a rubber impression roller and the image thus transferred to the web. The etched (printing) areas are divided into cells, the depths of which control the amount of ink deposited at any point and, hence, print density. Gravure is thus able to reproduce half-tones efficiently and is used for high quality work. Preparation of the etched metal cylinder is a long and costly process, however, and can only be justified by long runs.

Gravure inks are based on pigments dispersed in a resin solution and must be of low viscosity in order to fill the cells in the metal cylinder. For packaging work, where odour and taint may be important, low odour solvents, such as industrial alcohol, are normally used. The principle of gravure printing is shown in Figure 10.1.

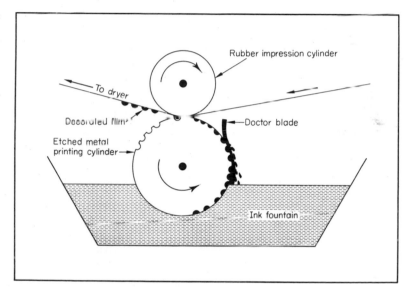

Figure 10.1 Principles of gravure printing

Flexography   This is the process most widely used for printing on reel-fed plastics films although it is also used for printing other reel-fed materials. The printing roller is covered with a flexible plate with the printing areas raised above the plate surface. The printing areas are inked by a roller rotating in an ink duct. The print is transferred to the paper or film at the nip between the printing roller and a steel impression cylinder. Very fine detail cannot be achieved but the flexible plates are relatively cheap and changeover times are short. Printing speeds are higher than for gravure.

Improved metering of the ink is achieved by the use of an engraved roller (anilox roller) in the ink duct (Figure 10.2). For film and foil printing, flexographic inks are similar to alcohol based gravure inks. For

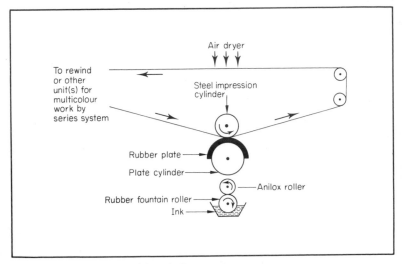

Figure 10.2 Flexographic printing process

paper and board, water based inks can be used, and these are normally odourless.

Letterpress   This process is also based on raised characters and can be flat-bed or rotary. In the flat-bed process, the bed carrying the printing plate reciprocates, passing first under a series of inking rollers, then back under an impression cylinder carrying the paper or board. It is a slow process because the bed is heavy and has to change direction twice for each impression. This is overcome in the rotary method which utilises curved printing plates clamped to a rotating cylinder. In packaging uses most letterpress inks are oil-based, drying by a combination of oxidation and absorption. Such inks are not suitable for non-absorptive materials such as plastics films, because set-off will occur between the layers in a reel.

Dry offset   As its name implies, this is not a direct process. Ink is transferred to a plate cylinder, with the printing areas in relief, and is then picked up by a rubber cylinder, called an offset blanket, and transferred to the material being printed (Figure 10.3). It is often used for the printing of plastics bottles. Inks are similar to those used for letter-press.

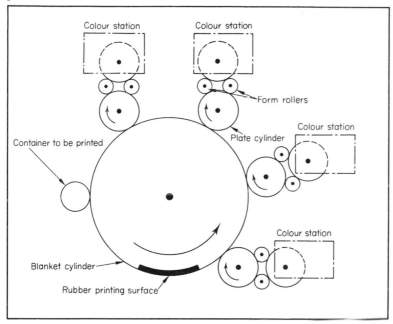

Figure 10.3 Dry offset printing process

Lithography For packaging this is an offset process and is rotary. A thin metal plate is chemically treated to produce ink-receptive and water receptive areas and then clamped around the printing cylinder. This rotates, first against a damping roller, next against inking rollers where the oil-based ink is repelled by the water film but accepted by the ink-receptive printing areas. The ink is transferred to an offset blanket, then to the material to be printed. The process is widely used for printing tinplate but is also used for cartons, labels and aluminium foil. Inks dry by oxidation or by oxidation and absorption.

Silk screen This is a stencil process, the screens being porous only in the printing areas. Ink is pushed through the screen by a rubber squeegee (Figure 10.4). The process is widely used for the printing of plastics bottles and for advertising material, where a heavy lay-down of ink is required.

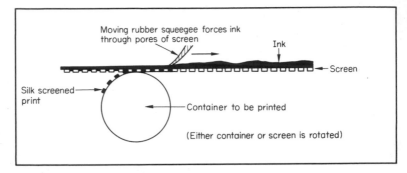

Figure 10.4 Silk screen printing process

Distribution

Distribution is a vital part of marketing and one which is very dependent on packaging. The main functions of distribution are those which relate to the handling, transporting and storing of goods.

Packaging and handling

The handling of goods is affected by packaging in many ways, the chief of which are:

1 Size of unit package
2 Shape of unit package
3 Rigidity or otherwise of the unit package
4 Number of unit packs collated in shipping packs
5 Any further consolidation of shipping packs into larger van-type 'containers'
6 The type of collation used for unit packs
7 Weight of pack

Size of unit pack This has a bearing on whether or not collation of the unit pack into larger shipping units is necessary. It may also dictate the type of handling used ie manual forklift truck, crane, etc.

Shape of unit pack This also has a bearing on the type of handling and on whether collation into larger units is necessary. In the case of manual handling of large containers, the shape can affect the type of mechanical shock to which it will be subjected. For example, a drum may easily be rolled along the ground whereas a heavy box is likely to be tipped end over end with an appreciable impact at each movement.

Rigidity of unit pack   This affects the ease of handling, the ease of palletisation and the type of outer container, if one is necessary for reasons of size and shape. A unit pack with low rigidity would normally have an outer container that is able to give adequate support and protection during handling.

Number of unit packs per shipping pack   This is often dictated by the requirements of the eventual sales outlet. The number chosen affects the size of the shipping pack and thus affects its handling.

Containerisation   This is the principle of consolidating a load of normal shipping packs or large unit packs into a large van-type container. These are made of steel, plywood, aluminium or plastics and are loaded at the factory on to a flat-bed lorry or freight car and fastened into place. Advantages include reduction in pilfering, simplification of paper work and reduced handling costs.

Type of collation used for unit packs   The major shipping pack for unit containers such as jars, cans and cartons, has been, for many years, the fibreboard outer case. This is easily handled, can be designed to protect the contents efficiently and is of reasonably low cost. In recent years, however, the concept of shrinkwrapping has become popular, and shrinkwraps are replacing fibreboard outer cases to an increasing extent. Shrinkwrapping consists in loosely wrapping the article or articles to be packed with a plastics film (usually polyethylene) which has been stretched under heat during its manufacture.

The loose wrap is passed through a hot air oven where the film shrinks to give a tight, contour wrap. Where cartons are shrinkwrapped it is usually sufficient to build a suitable stack of cartons and proceed as above. With cans, jars or bottles it is usually necessary first to collate a suitable number of containers on to a shallow fibreboard tray and then shrinkwrap the tray and contents. Two or more trays can be stacked together, if desired, and then shrinkwrapped.

Whole pallet loads can also be shrinkwrapped, giving more stable and weatherproof loads. Shrinkwrapping does not usually give as much mechanical protection to the contents as does a well-designed fibreboard case, but it is cheaper and has the psychological advantage that the contents can be seen to be fragile and are often handled more carefully. It is preferable to collate glass containers on a thermoformed tray in order to keep them out of contact with each other.

The coefficient of friction of the film is important if the goods are to be palletised and it is necessary to omit slip additives when making film for shrinkwrapping.

Weight of pack   The weight of the pack affects the type of handling it receives. As a rough guide, manual handling will be used for up to 40 pounds (18kg), and automatic handling for heavier packs.

Packaging and transport

Since transport costs for a product include an element of cost for the package, it is important to take into account the demands of transport when designing a package.

Shape is obviously important from a space utilisation point of view, but so is weight (especially in air transport). Fragility of the package itself is important because of factors such as vibration and sudden shock, either of which may often be encountered. In the case of shrinkwrapped articles, mentioned above, the main problem in transport is one of mixed loads. The transport of shrinkwrapped articles next to, say, metal jerricans or wooden crates can easily lead to puncturing and tearing of the plastics film with consequent collapse of the load. However, shrinkwrapping has another role to play in improving transportability of goods. This is its use in the overwrapping of whole pallet loads. A shrinkwrapped pallet is not only better protected against climatic hazards than a board or paper wrapped one, but is also more stable.

Finally, of course, better packaging means lower transport claims although there is a law of diminishing returns in this connection and expensive over-packaging should be avoided.

Packaging and storage

Storage at either end of the journey is often necessary and packaging can again play a useful role in reducing storage costs. To a certain extent, what has been said about handling and transport applies also to storage. With plastics it is necessary to stress one thing in particular with regard to their behaviour on stacking. Thermoplastics such as high density polyethylene and polypropylene when used for crates and tote boxes, are subject, to a greater or lesser extent, to cold flow or creep.

This means that a certain amount of distortion can occur over a period of time in the lower members of a stack. The phenomenon is strongly time dependent and is not necessarily shown up by short-term compression tests. Perfectly satisfactory results can be obtained, however, provided that the problem is recognised and the containers are correctly designed with adequate thickness of wall section at points of maximum stress. Alternatively, outer cases can be used which are stout enough to support the load.

The static part of storage is not the only thing to be considered in packaging design. Modern warehousing is a complex business and packaging must facilitate replacement and recovery of goods, as well as rapid movement of the goods through the warehouse. In addition to its job of protection, packaging can often be the means of overcoming problems due to an awkwardly shaped product.

The functions of a modern warehouse can be broken down into the following elements:

1 Unloading
2 Moving goods into storage
3 Moving goods from reserve to forward stocks
4 Making up orders
5 Shipping and delivery

Over the past years there has been a tendency towards single-storey warehouses. Efficient use of floor space in such a building necessitates high stacking (sometimes as high as 25-30 feet) and this puts a greater emphasis on package strength. Without the development of stronger packaging, the development of high stacks would not be possible and warehousing would be more expensive.

The efficient movement of goods within the warehouse, and ease of stacking to greater heights, both depend on the use of pallets which in turn are dependent on well designed packaging.

Conclusion

Packaging is an important marketing tool but like any other tool it needs to be kept in good condition.

The implications of this can best be illustrated by reference to the various sections discussed earlier.

Retail marketing trends

The interdependence of packaging and trends in retail marketing has already been stressed. It follows that such trends must be closely studied and the packaging implications analysed. In addition there must be an awareness of changes in social habits since these are often the initial cause of retailing changes. An increase in the number of working wives, for instance, brings in its train an increasing need for convenience packaging together with a need for convenience shopping; for example, supermarkets.

The increasing cost of labour leads to an increase in the use of vending machines at railway stations, industrial sites, institutions, etc. However good the product it will not achieve maximum sales unless its packaging fits retail requirements.

Design and company/brand image

The lesson here is simply that any up-dating of the company/brand image must be accompanied by up-dating of the package since the package is one of the main lines of communication with the consumer.

Sales promotion

A new sales promotion project will usually boost sales for a short time only but a longer term impact can often be achieved by tying this in with a packaging face-lift.

Advertising

It is important to ensure that packaging or advertising changes are not made independently, without consideration of the effect one may have on the other. A simple thing, such as a change in colour, may make it unrecognisable on black and white TV. This could happen because of a sudden lack of contrast due to the fact that two colours, distinguishable in themselves, both appear as grey on black and white TV.

Distribution

Distribution patterns should be checked regularly to ensure that some change has not rendered the product under- or over-packaged. One major change, of course, would be from home trade only to home trade plus export. Changes of a lesser magnitude can also have an effect.

11

Packaging and the Law

There is, unfortunately, no separate discernible branch of law or set of statutes which may be conveniently classified as 'Packaging Law'. It is necessary, when trying to find the law in any particular instance, to pursue a way through a maze of legislation on such subjects as Contract, the Sale of Goods Act, Weights and Measures laws, the Food and Drugs Act, Transport Legislation, Trade Marks and Design laws and many others. While there is no substitute for a good lawyer, this chapter is designed to give guidance on the most important penal statutes which affect packaging today. In addition to these penal statutes (available through HM Stationery Office) there is, of course, still in existence the historical body of common law to which the consumer may have recourse as a remedy for breach of contract or for negligence.

The legislation discussed in this chapter has been grouped for ease of reference, under the following headings:

1 Weights and Measures Act 1963
2 Food and Drugs Act 1955
3 Migration from packaging materials into food
4 Drugs, poisons and medicines

5 Transport and handling of dangerous goods.
6 Trade Description Act 1968.
7 Copyright, design, patents and trade marks.

Weights and Measures Act 1963

This Act is based on two general principles. One is that no one who sells goods by quantity shall deliver to the buyer a lesser quantity than is made out by the seller. 'Quantity' includes the determination of number as well as weight and 'deliver' means putting the buyer physically in possession of the goods. The other principle is that where goods are sold, or exposed or offered for sale by quantity, no one shall misrepresent the quantity or mislead either the buyer or the seller about it.

The Act requires a wide range of goods to be marked with their weights. Thus, the containers of all pre-packed foods must be marked with a statement of quantity. In the case of many basic foodstuffs (such as butter, tea, jam, sugar, etc), they may only be sold by retail in certain prescribed weights. This has a limiting effect on the type of packaging which can be used.

Schedules of the Act set out details of the goods which are subject to weight marking. These include soap, toilet preparations, scouring powders, detergents, seeds, fuel and distemper. The ways in which these requirements affect packaging are set out below:

1 The materials used for the package must be of a nature that ensures that the contents are indeed of the stated weight at the time of sale. The quantity and quality of the contents must also remain unaffected by storage conditions. This is important since it is often a good defence under the Act if the manufacturer can prove that his marked goods have been held for an unreasonable time at retail level. Thus, if the goods have not been interfered with or damaged but are still short weight then the manufacturer may be able to prove by the date coding on the pack that an unreasonable time has elapsed between packing and sale. This defence would fail if the package were of such a standard that it provided poor protection for even a short shelf-life.
2 Statutory declarations, such as a weight marking, form an obligatory part of the package design.
3 Regulations made under the Act decree the manner in which weight

marking, etc must be shown on the containers. For example, the normal rule is that the manufacturer cannot yet use metric terms alone but only in conjunction with the imperial terms. The only exception to this is in the field of perfumes and toilet preparations where metric terms may be stated in isolation.

4 There are certain regulations governing the size of lettering on the container used for pre-packed goods. This must be related to the greatest dimension of the container and it must be conspicuous and legible. The Act also lays down that the lettering must be in a contrasting colour to the background on which it is written. An exception to this is made in the case of embossed lettering which need not then be in a contrasting colour.

There is also a general exception to this in the case of foodstuffs. Here the packages do not yet have to be marked in any precisely defined dimension or character, provided that the lettering is conspicuous and legible.

'Pre-packed' is defined in the Act as meaning; 'made up in advance ready for retail sale in or on a container'.

Food and Drugs Act 1955

The main object of this Act is to ensure that no food which is injurious to health shall be sold for human consumption and infringements are punishable by fine or imprisonment or both. It also sets out to prevent false or misleading claims being made for food whether by means of the package or by advertisement. The Ministers concerned have wide powers to make regulations which amplify the objectives of the Act and many have in fact been made. Regulations having a bearing on packaging include those relating to the composition of foods, the use of particular ingredients in food, food hygiene and the labelling and advertisement of foods.

Labelling of Food Regulations

The most important regulations from the packaging point of view are the Labelling of Food Regulations 1970. These have been published but do not come into force until 1 January 1973 and will then revoke the Labelling of Food Regulations 1967. Among the provisions of interest to packaging management are the following:

1 All pre-packed foods have to show on the label the name of a responsible person such as the packer, labeller, or the person on whose behalf the food is labelled or packed. A registered trade mark may no longer be given as the alternative.

2 All pre-packed foods for retail sale have to bear the common or usual name, if one exists, (for example, custard powder) or an 'appropriate designation'. 'Appropriate designation' is defined as 'a name or description, or name and description sufficiently specific, in each case, to indicate to a prospective purchaser the true nature of the food to which it is applied'. Manufacturers may choose an appropriate designation for a food except where one has been laid down in the Act or in compositional regulations. Brand or coined names which are being used widely for all foods may still be used in conjunction with the appropriate designation. If they have been used for thirty years before 4 January 1971, they may be used on their own, provided this would not be misleading to the consumer.

3 Most pre-packed foods are required to carry on the labels a complete list of ingredients, including any additives, in descending order of weight with a heading, such as 'Ingredients'. Exemptions include biscuits, bread, butter, cheese, condensed and dried milk, coffee and coffee essence, ice-cream, flour and confectionery. However, for some exempted foods it is still necessary, if they contain additives, for the label to carry a special declaration such as, 'contains preservatives', or 'contains sulphur dioxide'.

4 The Regulations include requirements as to the height of the letters and the visual emphasis and prominence to be given to the common or usual name or appropriate designation, as compared with anything else appearing on the label. Minimum heights of lettering, related to the greatest dimensions of the container, will be required for the common or usual name, or appropriate designation, and the list of ingredients. The smallest letter on cans up to 120mm high (which includes most of the more commonly used cans) must be at least 2mm in height for the name or designation and 1mm high for ingredients. The list of ingredients must be next to the name of the food or within a surrounding line or on a contrasting panel so that the consumer can find it easily.

5 Labels and advertisements for dried and dehydrated foods have, as a

general rule, to include the word 'dried' or 'dehydrated' as part of the name or designation except where they are customarily sold in the dried state under names such as 'currants', 'raisins' or 'prunes'. Dry mixes must bear a statement of any ingredients, other than water, that have to be added to the mix in order to make up the food. There are also requirements covering the labelling and advertisement of foods that include in the name an indication that they have been flavoured by another food, and about the composition of foods sold as liqueur chocolates and shandy-type drinks. For example, liqueur chocolates must contain a liquid filling comprising a liqueur, spirit or fortified wine in a significant quantity. This automatically excludes cream filled chocolates flavoured with a liqueur. Shandy-type drinks must have a minimum proof spirit content of 1.5 per cent.

Colour in Food Regulations

These regulations lay down the colours which may be used in foodstuffs. If a prohibited colour should find its way into the food via the packaging material then the packaging manufacturer would be involved.

Hygiene

Regulations prescribing hygiene in food businesses have also been made under the Food and Drugs Act. For the purpose of these regulations, the handling of the food is taken to include the storage, packing and wrapping of food, and provision is made that no wrapping material shall come in contact with food which is liable to contaminate it. Also, only printed material designed exclusively for the purpose shall come into contact with the food.

Migration from packaging materials into foodstuffs

Although there is not, at present, any legislation in the UK on this subject it is included here because the subject is under consideration following publication of a report in 1970 by the Food Additives and Contaminants Committee on 'The leaching of substances from packaging materials into food'. This report is available from HMSO. Cognisance must also be taken of the fact that some other countries (notably the USA, France, Holland and Italy) do have legislation in the form of positive lists of permitted packaging additives.

In the case of plastics for food contact use, (including food packaging) the British Plastics Federation (in collaboration with the British Industrial Biological Research Association — BIBRA) have issued a code of practice as a guide to industry in choosing components for plastics packaging materials that are safe in contact with food.

Germany has 'recommendations' issued by a committee of government officials and industry members. These have no legal validity but are recognised by the German courts as defining technologically unavoidable migrants.

In general, the philosophy behind modern legislation on this subject is that the burden of proving safety in use is on the manufacturer or user of the packaging material. Materials whose safety in use have been demonstrated are then included in some 'positive' list of permitted materials.

Two factors are taken into account in assessing safety-in-use of a particular food packaging material.

1 The toxicity of the base materials and of any components added to achieve certain desired properties in the finished package or as an aid in processing.
2 The amount of each base material or additive which finds its way into the food by migration.

The results of any migration must not make the food unfit for human consumption. This means that the quantity of each component migrating into the daily food must be less than the 'acceptable daily intake' of that component, per person of standard weight (taken to be 60kg).

In some countries, the legislation also takes 'unfit' to cover undesirable changes in odour, flavour or colour. Such changes are extremely difficult to classify as acceptable or not acceptable.

Drugs, poisons and medicines

There are a number of statutes affecting medicines, poisons and dangerous drugs and they require precautions to be taken in the handling, labelling and sale of drugs, medicines and poisons. Failure to comply, whether or not damage or loss follows, leads to prosecution. Penalties may be fines or, in extreme cases imprisonment. Some of the more important from the packaging point of view are discussed below.

Dangerous Drugs Act 1951

This Act controls the manufacture, distribution and sale of certain dangerous drugs. Packaging is affected since the containers for those drugs listed in part two of the Dangerous Drugs Regulations 1953, must be marked with the particulars laid down in the regulations.

Pharmacy and Poisons Act 1933

This Act controls the manufacture, distribution and the sale of poisons. Regulations have been made under the Act which list the poisons subject to control, and the way in which they should be handled. There are many packaging implications. For example, subject to certain exemptions, no person may sell or supply a listed poison unless the container is labelled in the prescribed manner with the name of the poison, or in the event of a preparation containing poison as an ingredient the proportion of that poison. The name 'Poison' must also be on the label together with the name of the seller and the address from which the poison was sold.

Penalties may also be incurred if any poison is stored or sold, whether by retail or wholesale in a container which is not impervious to the poison and which is not sufficiently stout to prevent leakage arising from the ordinary risks of handling and transport. In the case of a liquid poison contained in a bottle of up to a prescribed capacity, (120 fluid ounces) the bottle must, with certain exemptions, be fluted vertically with ribs or grooves sufficiently obvious to be recognised by touch. The transportation of the poisons is also covered and it is unlawful to consign poisons for transport unless they are stoutly packed to avoid leakage. In addition, certain listed poisons may not be consigned for transport unless the outside of the package is labelled with the name of the poison and a notice saying that it must be kept separate from food and from empty containers in which food has been contained.

The Pharmacy and Medicines Act 1941

The containers in which medicines are packed have to be labelled with an appropriate designation of the medicine or else with a list of the ingredients and their quantities. This Act applies not only to medicines in the strict sense of the term but also to preparations which are not necessarily medicines but have curative claims made for them.

Medicines Act 1968

Although this Act has received Royal assent it has not yet come into operation. Under Part V of this Act the Ministers are empowered to make regulations concerning the labelling of containers and packages containing medicinal products. The regulations may cover identification, warning notices, safety, and ensuring that false information is not given.

Other regulations which can be made concern the material, strength, shape or pattern of containers and the colour, shape and distinctive markings on medicinal products.

Transport and handling of dangerous substances

In addition to specific legislation such as the Petroleum Acts and Regulations which lay down conditions for the manner and containers in which petroleum may be conveyed, stored and sold, there are certain legal obligations imposed by common law. Thus, the manufacturer has an obligation to warn those who may be exposed to risk when handling goods manufactured by him. Once dangerous chemicals have been dispatched from the point of manufacture they are liable to come into contact with many people. These include carriers, warehousemen, porters and even the general public. In addition, there are the manufacturer's employees and customers. All these should be warned of any dangers which could arise from handling the particular product.

Since 1953, the British chemical industry has operated a voluntary labelling scheme as a means of fulfilling its common law obligations. The scheme is described in a manual published by the Chemical Industries Association Limited and entitled, 'Marking containers of hazardous chemicals'. The layout of a standard label is shown in Figure 11.1. It contains:

1 The name of the product.
2 A signal word, giving the degree of danger. The words CAUTION, WARNING or DANGER denote increasing degrees of risk.
3 The type of hazard. This is outlined by the use of standard phrases which have precisely the same meaning on each occasion.
4 The precautions to be taken — again using standard terms.
5 Appropriate action to be taken, and first aid to be given in case of accident.

D–D Soil Fumigant

WARNING

A volatile, inflammable liquid, hazardous if swallowed, inhaled or absorbed through the skin. If exposed to the skin wash thoroughly with plenty of soap and water especially before eating or smoking. Do not use contaminated clothing until it has been laundered. If swallowed drink large amounts of warm water and induce vomiting with an emetic of a tablespoon of salt in a glass of warm water; keep the subject prone and quiet; call a doctor

KEEP AWAY FROM HEAT, NAKED FLAMES AND SPARKS
IN CASE OF FIRE USE SAND OR EARTH

Figure 11.1   Example of warning label

The Petroleum Acts mentioned contain provision for labelling the prescribed containers with suitable warnings. Gas cylinders are also subject to similar requirements. The Inflammable Substances (Conveyance by Road) Labelling Regulations 1968 which were made under the Petroleum Acts, require vehicles carrying certain inflammable materials with a flash point below 73°F (23°C) to carry danger warnings. The containers must also be marked in accordance with the requirements of the schedules to the regulations.

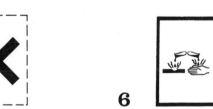

1 Black bomb on orange ground
—Liable to explosion

2 Black flame on orange ground
—Danger of fire

3 Flame over a circle, black on orange ground
—Inflammable substance

4 Black death's head on orange ground
Toxic substance

5 Black St. Andrew's cross on orange ground
—Harmful substance

6 Liquid dripping from a test-tube onto a plate and from another test-tube onto a hand, black on orange ground
—Corrosive substance

Figure 11.2 Pictorial markings for hazard labels

Pictorial markings for dangerous goods

The United Nations have drawn up a classification and labelling scheme for dangerous goods. These are important because they form the basis for many national regulations. The UN scheme has also been incorporated into the International Maritime Dangerous Goods Code compiled by the Inter-Governmental Maritime Consultatives Organisation (IMCO). The UN labels are also reproduced in the UK Report of the Standing Advisory Committee on the Carriage of Dangerous Goods in Ships, commonly known as the 'Blue Book', and issued by the Department of Trade and Industry.

The Blue Book classifies products according to the types of hazards, sets out the packages which are acceptable and in some cases sets out the maximum quantity of product which may be packed in each type of package. It also details the stowage restrictions and the symbolic hazard labels to be used. Typical pictorial markings for hazard labels are shown in Figure 11.2.

Sea Transport

The Blue Book, and the IMCO Dangerous Goods Code have already been mentioned above. Note must also be taken of The Merchant Shipping (Dangerous Goods) Rules 1965, which apply to all British ships registered in the United Kingdom, and to all other ships while they are loading cargo within any port in the United Kingdom or within its territorial waters. These rules specify that it shall be unlawful for dangerous goods to be taken on board unless the shipper has made a declaration, in writing, of the identity of the goods, the class of hazard to which they belong, and that they are properly packed, marked and labelled in accordance with the rules.

Air transport

The International Air Transport Association (IATA) has published regulations (The Restricted Articles Registration) which also classifies products by hazards, indicating which are not acceptable and which are acceptable only in cargo aircraft. They also specify a maximum weight of product which can be carried, the construction of the package and the tests it must be capable of surviving.

Road transport

The Home Office is empowered to issue regulations to cover the transport of hazardous goods by road in the UK, under The Petroleum (Consolidation) Act 1928. Regulations exist covering construction and loading of vehicles and the marking of both transit package and the vehicles with internationally agreed hazard labels of the type shown in Figure 11.2. As at December 1971, however, no packaging requirements had yet been stipulated.

Rail transport

The transport of hazardous goods by rail is governed in the UK by British Rail's conditions of acceptance. These are set out in their publication 'Dangerous Goods by Freight Train and by Passenger Train'. Dangerous products are classified by hazard and the packages which will be accepted are specified. The international labels (as typified in Figure 11.2) have been adopted and combined in a single label with BR's own loading instruction labels.

Trade Descriptions Act 1968

This Act replaced the Merchandise Marks Act but is of much wider application. It also creates an authority charged with a positive obligation for enforcing the new law, namely, the local Weights & Measures Authorities.

Section 1 of this Act makes it a criminal offence either to apply a false trade description to goods, or to sell or to have for supply goods to which a false description has already been applied. The first offence is the one most likely to apply to packaging and it has already been held that a misleading package can constitute a misleading description under the Act. A misleading package may be one which imputes a quality not present in the goods contained in it, or one which infers a much greater quantity of product than is, in fact, present.

Section 2 contains a prohibition of false statements about the size of goods and this has already led to a complaint under the Act in respect of packaging. The case concerned a manufacturer of children's interlocking plastic bricks which were sold in a retail store and displayed in the window. The package was arranged so that a close-packed section of the bricks could be seen through a cellulose film window in the carton. However, the area of the remainder of the box, obscured by cardboard, was in fact mostly filled by cardboard. An assessment made on the boxes showed that there was about 38 per cent less in quantity or size of bricks than would be expected from the outside packaging.

A similar case arose with a stain remover packed in a tube. The tube was packed in a carton and held rigid by various shaped pieces of cardboard. It was held that potential customers, having no visible indication of the contents, were led to believe that they were getting more of the stain remover than in fact they were. Complaints have also been made under the Act against dispensing mechanisms supplied with certain types of adhesive tape, saying that the dispenser gives an exaggerated idea of the circumference of the roll of tape contained in it.

The printing on the packaging can also lead to prosecution under the Trade Descriptions Act, if it is held to be misleading as to the quality or quantity of the contents. An example here is the case of a cake mix which showed a picture of the finished cake on the package, presumably as it would look when baked by the housewife. Beside the cake there were depicted certain table furnishings, knives, forks and spoons. The size of these was such as to give an exaggerated idea of the size of the cake. A similar package for a harvest pie was illustrated with a picture which gave an extremely exaggerated idea of the number of bilberries likely to be found inside the pie. No prosecution was made but the manufacturers and suppliers concerned gave an undertaking that future packages would carry illustrations more in line with the contents. An interesting sidelight on this case was the fact that the manufacturers agreed that until the supply of cartons was exhausted, they would ensure that an overprinted flash would be added at the point of retail sale, to obliterate the faulty representation.

One point of interest as far as the packaging supplier is concerned is to what extent he is responsible when he has produced packaging to the specification of a manufacturer or supplier of goods. The Act does not only apply to sellers of the goods because there is a provision in Section 1 that it applies to people acting in the course of trade or business. Therefore, the man producing the packaging containing a false trade description can be guilty of an offence. However, there are certain defences for those who have innocently been asked to supply packaging or advertising material which is later found to be inconsistent with the contents. First, they should not have known of the falsity of what was said on the package and secondly, they should not have been able, with reasonable diligence, to have discovered the error of what was produced.

Finally, a point of utmost importance to managers. Section 20 of the Act provides that, apart from the employee responsible for a particular mis-description, responsibility is extended to anyone who has a degree of responsibility for what has occurred. This includes directors of the company who had overall management responsibility for particular activities, senior employees, managers or the company secretary.

Copyright, design, patents and trade marks

These four subjects are conveniently considered together since they are all concerned with protecting private rights of property. Each of them is covered by a separate statute and each is designed to protect the designer, artist, inventor or writer against the unauthorised use of his own work. In general, each statute establishes the nature of the works which may be protected, the forms of protection given and the remedies available to the owner in the event of infringement of his rights. These statutes will now be considered separately.

The Copyright Act 1956

Copyright law in general is designed to give protection to a writer or artist against the unlawful reproduction of his work. Copyright cannot be given to an idea, however, and it is necessary for the work to exist in a concrete form before protection is effective. No registration is necessary; copyright comes into existence when the work is actually published. Copyright affects packaging matters in two different ways:

1 In the protection of the company's printed matter on containers or advertising material, and
2 In an awareness of the possibility of copyright held by others.

The first is not of very great importance although in some cases the written matter on a container may be capable of protection. Details of a trade competition may be capable of copyright protection, for example, but it has been held that copyright does not exist in the matter of advertising slogans.

Copyright may well be of importance in the case of a company wishing to reproduce written or photographic matter as advertising matter or on a container. For example, the right to actually publish a letter of testimonial does not automatically belong to the recipient. The permission of the writer of the letter must be obtained first. The reproduction of portraits or photographs is also fraught with pitfalls. A photograph may be taken of a public place, and used on a container or in advertising matter. Care must be taken if any person or persons also appear in the photograph since if their consent has not been obtained they are at liberty to complain or bring an action on the grounds that they could be thought to endorse the product. A way round this problem is to make their faces unrecognisable in the reproduction.

Another problem arises in connection with international copyright. A container may have been designed for the UK and illustrated with a photograph or drawing. In the case of expensive modelling fees it may well have been decided to buy only the UK rights, especially if the product was intended for the UK market at the time. If it is later decided to export the package then full rights should be obtained. It is, of course, quite conceivable that the rights in the photograph for some other country have been disposed of in the meantime by the model or his/her agent.

In general, copyright exists during the lifetime of the author and then for a period of fifty years after his death.

The Registered Design Act 1949

Industrial designs may also be copyrighted but here the protection is secured by registration at the Patents Office under the above Act. There is a prescribed procedure and a fee must be paid. It differs, therefore, from the sort of copyright discussed earlier. The time during which protection is afforded is also different. The initial period is five years and this can be extended at five-yearly intervals up to a total of fifteen years.

The Act defines design as: 'features of shape, configuration, pattern or ornament applied to an article by any industrial process or means, being features which in the finished article, appeal to and are judged solely by the eye, but do not include a method or principle of construction or features of shape, or configuration which are dictated solely by the function to which the article to be made in that shape or configuration has to perform'. From this it will be seen that a design must have reference to an article. Design registration cannot exist solely as an idea.

Up to 1968 the registration of a design was of paramount importance. If the design was used commercially but not registered the designer could not register it subsequently. An additional Act, the Copyright Act of 1968, has altered the position, however, and in the circumstances outlined above the designer has copyright protection for a period of fifteen years from the date the industrial articles were first sold.

Patents Acts 1949-1961

The granting of letters patent for an invention confers on the person or company granted such letters patent, the sole right to manufacture, use or sell the invention for a period of sixteen years. There are very many areas where patents are of importance in packaging. In addition to novel packages or novel ways of making packages there is a vast field covering improvements in packaging and filling equipment. Many patents also exist for tamperproof closures or for closures incorporating novel dispensing features. A written application for a patent has to be filed at the Patents Office and must be accompanied by either a provisional or a complete specification. Early application is essential as this establishes a priority date. Prior publication or use of the invention invalidates any resultant patent, although use is allowed for the purpose of '... reasonable trial and experiment'.

The complete specification can be filed with the application or within fifteen months of filing the provisional specification, and must contain:

1 A full and clear description of the invention.
2 Instructions whereby a skilled man could make the article or work the process.
3 A definition of the scope of the invention claimed.
4 Where necessary, drawings to illustrate 1 and 2.

If the complete specification is accepted, it is published and is open to inspection. For a period of three months anyone interested can register opposition to the granting of a patent. If no opposition is put forward, or if the opposition fails, letters patent are granted to the applicant in respect of his invention and it can be described as patented. The patent is granted for an initial period of four years from the date of filing of the complete specification but can be renewed annually on payment of fees which increase in amount each year.

Trade Marks Act 1938

Trade marks are symbols or names by which the customer identifies the products he or she buys. They are, therefore, an important factor in modern marketing and packaging.

Trade mark rights may be acquired by established use and are protected under the common law principle that no-one is entitled to pass-off his business or his products as, or for, those of another. The mark must, of course, be so well-known as to identify the goods in the market and this may not be so until the lapse of a considerable period of time. If it is desired to use a new trade mark then it can be registered under the Trade Marks Act and this confers upon the owner immediate and exclusive rights to the use of the goods for which it is registered. It also entitles him to sue for any infringement. It should be noted, however, that the same trade mark can sometimes be used by two different companies if it is applied to two different classes of goods.

Registration of a trade mark requires that it be distinctive or capable of distinguishing the owner's goods. It is not sufficient to use words which simply describe the quality or character of the goods to which they are applied, (for example, light, lovely, etc) nor to use emblems or devices which are merely pictures of the goods. It is important to note that if the owner of a registered trade mark does not take action against anyone who infringes that mark then he may lose the mark himself. This is in contrast with the laws governing copyright, design or patents. After a trade mark has been granted registration it continues in force for seven years. It can be renewed indefinitely for periods of fourteen years at a time.

Conclusions

Even the brief résumé given in this chapter shows quite clearly the importance of legislation to packaging management. The management approach can be summed up as follows:

1 Keep up to date on legislation likely to affect packaging.
2 Check new packs with legal experts.
3 Take care to protect trade marks.
4 Keep a check on competitors' activities.

Suggested further reading

F.A. Paine (editor). *Packaging and the Law.* London: Butterworth.
J.H. de Wilde and L.L. Katan. *Food Packaging and Health: Migration and Legislation.* London: Institute of Packaging.

12

Solid Waste Management and Packaging

The packaging industry is now under heavy attack by environmentalists and certain consumer groups on the grounds that it contributes appreciably to environmental pollution and is a major waste of the Earth's resources.

In the UK, for example, organisations such as Friends of the Earth have been campaigning against non-returnable packaging. In the USA this has gone further still and at one stage New York city imposed a two cent tax on plastics bottles. This was later overruled by the State Supreme Court Judge on the grounds that the tax was unfair and discriminatory.

The problem as far as management is concerned can be divided into two separate parts.

1 Design; the company's packaging should be designed with a view to avoiding over-packaging and aiding the problem of eventual disposability.
2 Public relations; much of the criticism is ill-informed and the problem should be put in its proper perspective to the public and to the relevant national and local government agencies.

Design

Design is an important factor in the related fields of over-packaging, disposability and recycling, and should be given due attention.

Over-packaging

Over-packaging is one of the targets of consumer groups and is often bracketed with the complaint of 'deceptive packaging'. The charge of over-packaging is often undeserved since few consumer groups are fully aware of the necessity for protection of the product. However, it is sometimes justified and a fresh look may be necessary at existing packaging, particularly if it is old-established, since changes in product, retailing methods, distribution methods, and storage, may have rendered the package over-protective.

Presumably if changes in the above factors had rendered the package under-protective this would have been detected by an increase in customer complaints or transport claims.

The question is not always one of over-packaging in the normally accepted sense. It may well be possible to reduce the amount of packaging material by a change in design. A reduction in a fibreboard outer case of a few square centimetres may only amount to a few grams per case but this could mean several tonnes less material to dispose of in the course of a year.

Savings of this type, multiplied by the number of different manufacturers of fibreboard cases, could make an appreciable contribution to lightening the load on waste disposal facilities and give appreciable cost savings. This type of argument makes shrinkwrapping (see Chapter Ten; 'Packaging as a Marketing Tool') an even more attractive proposition compared with a fibreboard case since it reduces the amount of packaging waste considerably. Apart from helping to solve the problem of packaging litter and waste, the reduction in packaging will also bring considerable cost savings.

Non-returnable packaging

One of the factors which has led to a great increase in the amount of packaging litter and waste is the trend towards the use of one-trip packaging. It is appreciated that there are often very good reasons for

227

the change from returnable to one-trip packaging but it is worthwhile considering each case strictly on its merits. In many instances, of course, the use of a so-called returnable container is no guarantee that it will, in reality be returned. Since the returnable container will be heavier than the equivalent one-trip container it is obvious that more of the Earth's material resources are being used up by the employment of 'returnable' containers in these circumstances. The other factor governing the choice of returnable or one-trip containers is that of consumer demand. There is a growing demand for one-trip packaging on the grounds of convenience to the consumer, a fact which is borne out by the unwillingness (already mentioned) to return even 'returnable bottles'.

The position is different in industrial packaging and the use of returnable packaging here is increasing. Manufacturing plants are finding it increasingly difficult to dispose of the packaging which brings in their raw materials, parts and assemblies, and they are, therefore, bringing pressure to bear on their suppliers to either cut down on packaging or introduce returnable packaging.

The economic picture, here, is not clear cut. With metal drums, for example, reconditioning costs can be quite high and there are also administrative costs to be considered. These can sometimes outweigh savings due to multi-trippage. Finally, there is the question of hygiene. Whenever food or pharmaceutical containers are re-used there is a risk of cross-infection, or of tainting, which is avoided by the use of one-trip containers.

Disposability

The term 'disposability' is taken here to mean any way in which the packaging material can be eliminated or converted at the end of its useful life as a package. There are, basically, three methods of achieving this.

Re-cycling   Re-cycling of packaging materials in this context is taken to mean the use of the discarded package to produce new packaging. This is not a new concept with some packaging materials, particularly with paper and fibreboard. Aluminium too has a long history of recycling; witness the collection of milk bottle tops and aluminium foil from cigarette packets. The steel from tinplate cans can also be re-cycled although the economics are not always quite so favourable.

228

The glass industry regularly uses broken glass (or 'cullet') as part of the mix for glass making. Cullet, in part, hastens the melting of a batch and so performs a useful technical purpose in addition to an economic one. The big question mark at present is against the re-cycling of plastics. One problem is the sorting of plastics waste. The different plastics used in packaging such as polyethylene, polystyrene, PVC, nylon, etc are very different from each other in their properties and could not be used in the form of a mixture for moulding into bottles or extruding into film.

Mixed plastics waste can be shredded and used in non-packaging applications but this will be dealt with later. The problem is complicated by the fact that even within one plastics type such as low-density polyethylene there are widely different grades used for different processes, for example, film extrusion, blow moulding and injection moulding. While not making the re-cycling of plastics impossible it does render the process uneconomic in most instances.

One possible exception could be where there is a large, easily segregated, market for one grade of one particular plastic. The archetype of this would be the milk packaging market if this went over to a standard plastics bottle. Such standard bottles could be segregated by the housewife and collected for re-cycling in the knowledge that only one grade, of one plastics species, was present.

There could still be a problem due to residual milk and other contamination but economic solutions could probably be found. Mixed plastics waste can be used in less critical packaging applications, one example being a plastics pallet, made by chipping scrap plastics which are heated and rolled under pressure. One of the problems with plastics waste is its large volume-to-weight ratio which makes it costly to transport. Paper can be pressed and baled, metal containers can be crushed and baled and glass can be easily broken and pulverised. However, equipment has been developed recently for shredding and compacting plastics waste and it can then be filled into sacks and transported in a much smaller space.

Non-packaging applications   There have been many approaches to this problem, some of them successful, although there are not yet any large-scale commercial outlets for waste packaging in non-packaging uses.

Waste glass, for instance, has been used as a filler in asphalt road making materials, under the name, glasphalt. The material has shown great promise in test strips of road laid down in various areas in the USA. Concrete has also been used as a matrix for waste packaging materials, notably, shredded polyethylene. Projects are still in the research stage but appear to offer an economic way of utilising waste polyethylene, and of producing a lighter-than-normal concrete.

Shredded plastics and partially granulated foams such as expanded polystyrene, have been used as heat insulating in-fill for building panels and the idea could even be extended to cushioning or heat insulation of transit packaging. Shredded paper and cellulose film, of course, have long been used as cushioning materials but have usually been obtained from the trim generated on slitting and reeling equipment. This is quite clean and it is doubtful whether collected waste paper could be utilised, except in certain non-critical applications.

A novel re-use of metal containers in Libya was that of beer cans discarded by a nearby American air base. These were used as containers for tree seedlings which were filled with earth and then buried in the desert. A slight amount of moisture was deposited in this particular area as dew. This would normally have been lost by absorption in the ground but the presence of the beer can around the seedling collected enough moisture for growth. Applications like this are, of course, specialised and not likely to make a large inroad into the overall problem. Nevertheless, they provide ideas for the design of packaging which can have a secondary use.

Finally, there has been a novel re-use in Germany of granulated expanded polystyrene waste as a soil conditioner. In addition to acting as a means of aerating heavy soils it is claimed that its use under sports fields, running tracks, etc gives a more resilient surface which is easier on the feet.

Waste disposal   In spite of what has been said above concerning re-use of packaging waste it seems certain that normal waste disposal methods will account for the major proportion for some time to come. In essence there are three methods of disposing of waste matter, namely, sinking it in the seas and oceans, burying it under the earth's surface, or incinerating it with subsequent dispersal of the resultant gases into the atmosphere. All three have been practised for as long as man has found it necessary to remove unwanted material from his immediate environment but the main methods used for disposing of household, factory or institution wastes are:

1 Land in-fill and
2 Incineration.

Land in-fill: In this method the refuse from the collection vehicle is transferred into the 'tip' where it is then covered with a layer of soil. Under these conditions the mixed waste gradually settles under the influence of various processes such as rusting, microbial breakdown, oxidation or other chemical action. The tipping area eventually settles and becomes stable, in terms of its density, water retention and drainage characteristics, and it is eventually restored to useful land.

This settlement and stabilisation of the tip is rendered more difficult if large quantities of plastics are present, because they do not easily rot or decay. Drainage is affected, for example, by the presence of plastics films which act as barriers, while plastics bottles can cause cavities in the tip structure. If the number of cavities is large the tip may not be suitable for supporting roads or buildings but it might be useful as pasture or for landscaping. The effect is not likely to be serious where plastics constitute less than about 6 per cent of the total tipped refuse. This figure is not expected to be reached until after 1980 according to figures published for the UK (*The composition, yield and analysis of domestic refuse* — A.E. Higginson — Institute of Public Cleansing, 73rd Annual Conference, June 1971) and for Germany (*Plastics Waste and litter* — Dr. ing. J.J.P. Staudinger — Society of Chemical Industry 1970).

Mechanical pulverisation or compaction of the plastics waste will aid its incorporation into the structure of the tip and will extend the usefulness of this method of waste disposal for some years past the deadline given above.

The criteria for suitability of a package for use in land-fill operations are initial density, ease of compaction (increasing the density) and degradability. The higher the density, the less tip space will be wasted so that a package with a high initial density or one which can easily be compacted to a higher density is to be preferred. Similarly, a degradable package will aid tip settlement.

Glass is easily compacted by pulverisation techniques using hammer mills, while paper and board can be shredded and baled, using ram operated compactors. Metal containers can also be compressed and baled. Some plastics packages are too resilient for straight compression and must be shredded prior to compaction. While shredding should be

231

no problem at the tip, there may well be adverse criticism, from the housewife, of packages which take up too much space in the dustbin.

One way of improving compressability of plastics containers is to design them with ribbing or diaphragms which would give adequate strength when full but which would aid crushing when empty. The use of ribbed bottles is common in Europe (eg PVC bottles for salad oils, wine etc). The question of reducing bulk is increasing in importance as suitable in-fill sites become harder to find and the rubbish must be conveyed for longer distances.

Incineration: As at 1971 this accounts for about 10 per cent of waste disposal but the use of incinerators is rising and is expected to reach 20 per cent by 1980. One of the advantages of incineration over land-fill is that useful energy can be obtained either directly, as heat, or by conversion to electricity. The heat can be utilised in communal central heating schemes while the electricity can be fed into the national grid system.

Plastics have a high heat value and are easily handled by the newer incineration plants now being built. Some of the factors to be taken into account when assessing the suitability of a package for disposal by incineration are:

1 Its rate of burning.
2 Its heating value.
3 The amount and composition of the residue after incineration.
4 The gases generated.
5 Corrosion or other damage to the incinerator.

The question of which gases are generated is important because of the possibility of gaseous pollution of the atmosphere.

Plastics, being relatively new materials, have come under fire on this account but in fact, with efficient incineration, most plastics should burn completely to give mainly carbon dioxide. This gas is the same as that exhaled by animals and utilised by plant life, which then gives off oxygen. PVC, however, will yield hydrogen chloride (HCl) which is unpleasant (though not so dangerous as sulphur dioxide which is present in the incinerator gases through other causes). However, if concentrations of either of these gases reach dangerous proportions they can be removed by wet-scrubbing in the chimney stack. This technique is in common use in modern chemical plants.

Degradable packaging   The concept of packages which will degrade easily when discarded is one which promises advantages, whether the package ends up on a disposal tip or is thrown away to become litter. It does, however, have disadvantages. Some packages do degrade when exposed to the weather. Tinplate eventually rusts away, though not quickly, while paper rots away at a comparatively fast rate. Glass is not a large component of litter but where it occurs it can be dangerous either through cuts or by causing grass fires by focussing the sun's rays. The materials causing the most concern are plastics as they will normally take anything from two to five years to degrade in the open air. The inertness and durability of plastics are, of course, two of their strong points when they are considered as packaging materials and no easy way has yet been found to make them fully satisfactory during use and easily destructible after use.

One suggestion which has been made is to render the plastics bio-degradable, for example, degradable by biological action, such as bacterial attack. This could have extremely undesirable consequences if a successful bacterial strain were developed which lived on plastics. Since, to be successful, they would have to be capable of living in the earth, they could well attack plastics waterpipes or underground cables, with devastating results.

Another approach is to utilise certain plastics which are water soluble. A bottle can be formed which consists of a sandwich, with a water soluble plastic at the centre. When the bottle is empty the outer coating is scratched before the bottle is discarded. The water soluble plastic is then exposed and will dissolve in damp conditions. This still leaves the thin insoluble coating material, however. One such bottle has been developed by Ilikon Corporation of Natick, Mass., USA and is based on hydroxypropyl cellulose.

The action of sunlight has also been suggested as a trigger for causing the degradation of plastics. Many plastics are susceptible to degradation by ultra-violet (UV) light and, in fact, UV stabilisers are added to plastics likely to be used outdoors to any extent, in order to prolong their life. Leaving out these stabilisers and replacing them with additives which cause breakdown of the plastics in the presence of UV light, has been suggested as a means of speeding the breakdown of plastics on being discarded.

The danger in this approach lies in controlling the exact time at which the breakdown begins. Premature degradation, while the bottle is in use, would be disastrous. The advantages too, become more nebulous on closer examination. The bottles or film on the disposal tip are soon covered either by earth or by a fresh load of rubbish and there may be only a short exposure to sunlight. The litter problem may be partially solved by this approach but at present the contribution of plastics to the litter problem is small.

Finally, there will be the problem of obtaining approval for the additives used when the plastics are to be used in food packaging. The other disadvantage is that the plastics are then lost and cannot be recycled.

Public relations

Although there is no doubt that packaging management should do all in its power to minimise the litter and waste disposal problem by due attention to its own packaging, there remains the fact that many of the problems have been grossly over-stated and, indeed, exploited by politically motivated groups. An efficient public relations campaign should be maintained, therefore, to put the subject in its proper perspective. Some ammunition for such a campaign is given below.

Litter

The charge levelled against packaging is that it is the main cause of litter. The volume of packaging litter is said to be large because:

1 There is a vast amount of over-packaging; packaging for which there is no technical justification and which the consumer does not want.
2 Non-returnable packaging is growing and is only being developed because the manufacturers want it, not because of consumer demand.
3 Ordinary waste disposal methods cannot cope with the newer forms of packaging based on plastics. This is a wide subject and will be dealt with later in detail.

The general charge that packaging is the main cause of litter is difficult to refute but it can be put into perspective more easily if we consider town and country litter separately. The only way to measure town litter is to consider the composition of the street cleaners' trolleys. This is not normally done but the results of one such exercise revealed the following:

Leaves:	63% by weight
Paper:	3.7% by weight
Metal:	2% by weight
Plastics:	0.5% by weight
Glass:	0.2% by weight
Wood:	1.3% by weight
Bricks and stones.	2.0% by weight
Sweepings:	27.3% by weight

(sweepings defined as sand, gravel, etc passing through ½in. mesh).

The high proportion of leaves is accounted for by the fact that the investigation was carried out in the autumn. If the leaves are ignored the percentage composition of the remainder is:

Paper:	10% by weight
Metal:	5.4% by weight
Plastics:	1.4% by weight
Glass:	0.5% by weight
Wood:	3.6% by weight
Bricks and stones:	5.4% by weight
Sweepings:	73.7% by weight

It will be seen that packaging does not contribute a large percentage by weight even if all the paper is taken to be packaging (and this is unlikely, since there will be an appreciable quantity of newspapers, hand-bills, etc). Unfortunately, packaging is a larger percentage by volume and is, therefore, easily seen.

Countryside litter is even more difficult to quantify. However, the results of an intensive clean-up of one particular area of countryside yielded 120 tons of litter. The make-up was not quantified but the contents of the haul included oil drums, bedsteads, paper, cars and household items. The lack of packaging items among the major constituents is significant. Dealing now with the separate items mentioned earlier the following information is relevant.

Over-packaging A great deal of the so-called over-packaging complained of by consumer groups is necessary for protection from either chemical or mechanical damage. This fact is worth putting over strongly in each case. Packaging is also often necessary for cheaper handling of goods, as in supermarkets. There remain two main types of packages which are pointed out as examples of over-packaging. One is the gift wrapped or highly ornamental package, while the other is the multi-pack of, say, six canned drinks in a board or plastic film overwrap.

Ornamental or gift wrapping may certainly not be needed for protection but it is required by the consumer in most cases. For example, very few women would buy a cheap glass bottle, simply labelled, and containing a perfume, toilet water or beauty cream. In many cases the pack is part of the product. (See Chapter Ten; 'Packaging as a Marketing Tool').

The multi-pack, too, can be defended, this time on the grounds of convenience in handling. A point quite often overlooked is that over-packaging reduces the manufacturer's profits and if his packaging was excessive he would soon be undercut by cheaper packaged competitive products. In addition, when market trials are carried out, the reactions obtained obviously refer to the product and package together so that an unacceptable package is unlikely to survive.

Non-returnable packaging This is a particular target of ill-informed criticism. One of the first facts which tend to make nonsense of attacks is that the general public is not particularly anxious to return even so-called returnable containers. The figure for return of 'returnable' beer bottles in the USA for example, has been falling over the past few years from about 30 to 40 per cent down to 5 to 10 per cent. This is partly due to increasing affluence during the same period and partly to the demand for one-trip containers as a convenience feature. The glass milk bottle is one example, at least in the UK, of a successful multi-trip container but here, of course, the consumer is put to no inconvenience but merely has to put the empty bottles on the doorstep. Even here there are wide variations in trippage, from four to forty per bottle, according to the region.

There is another argument against returnable packaging in the question of hygiene. The re-use of a container involves a great deal of money and labour in washing and sterilising the returned container and even then there is a risk of contamination. The manufacturer of a food or pharmaceutical product has no control over the use to which the consumer may put the container before it is eventually returned and it is impossible to inspect every one before refilling.

There will always be the risk of some taint or biological contamination remaining unaffected by the level of washing and sterilisation normally practised, especially with plastics bottles.

Waste disposal problems

Some of the arguments have already been dealt with earlier in the section on 'Design'. Thus, there could be problems with land-fill methods of disposal should the level of plastics in the total waste rise above 5 to 6 per cent, but this is not likely until about 1980. Even here, it will still be possible to use the land, after tipping has finished, for less critical uses than as building land. The problem of void formation in the tip can also be much reduced by shredding or by some other method of pulverising the plastics prior to burial. Machines capable of achieving this are already on the market.

Incineration of plastics is usually critical on two counts; difficulties with the actual incineration (clogging of fire bars, corrosion, etc) and pollution of the atmosphere by the resultant gaseous emission. The incineration difficulties refer to the older incinerators which were not designed to deal with plastics, but even here the difficulties have been exaggerated. However, the newer incineration plants, such as those at Edmonton (London) and South Shields (County Durham) are fully capable of handling plastics and the presence of plastics in the mix is said to aid combustion, especially when the mixture is damp. The development of these newer incineration plants has not been rendered necessary purely on account of plastics. One of the features of these new plants is a more efficient utilisation of the incoming waste including better heat utilisation and less emission of noxious gases and a more efficient use of the by-products.

Pollution of the atmosphere through the burning of plastics has also been exaggerated. Modern plants are more efficient and convert carbonaceous waste, such as plastics, to carbon dioxide and can prevent the escape of gases such as hydrogen chloride which is given off from burning PVC.

The economic advantages of burning packaging waste in general, also need stressing. Heat, for district central heating schemes, or steamraising for electricity generating plant, are two methods of utilising the potential energy of packaging waste.

Some of the problems of waste disposal could be alleviated if a means of preliminary sorting could be carried out by the housewife. This is by no means impossible. During World War II the housewife was quite accustomed to sort her household waste, separating paper and vegetable waste (for the council pig farms) from the remainder. It needs little more work to separate the glass, metal, plastics and paper and some financial incentive could be provided.

The problem of the volume of household waste as related to the capacity of the household dustbin is also not insoluble. Small shredders and crushers suitable for household use are already available and these can reduce household waste to about 10 per cent of its original volume. Once again, the final savings to the local authority might well allow some financial incentive to be given to the householder.

At present, the waste is usually crushed or compacted at the tip or by the collecting dust cart.

Use of the Earth's resources

The allegation that packaging is a waste of the Earth's resources is another popular anti-packaging argument used by environmentalists. It is most conveniently put into perspective by considering the various packaging materials in turn.

Glass The major component of a glass mix (nearly 75 per cent) is sand. Admittedly, not any type of sand is suitable for producing clear, water-white glass, nevertheless, there is still a large quantity available. If any shortage of high quality sand should develop, there is an almost inexhaustible supply of lower quality sand available. To this can be added the fact that a fairly high percentage of broken glass (already 10 to 30 per cent) can be added to the mix.

Aluminium Aluminium occurs very widely in the Earth's crust and it can also be reclaimed.

Paper/board Fresh trees are being planted each year and this is expected to keep pace with demand in future. Re-use of waste paper and board is already being carried out on a large scale and could be increased if segregation of paper waste was practised more widely. There are problems where plastics and paper are laminated but the problems of separation are being solved.

Plastics Much play is made with figures for oil reserves which purport to show that oil will run dry within the next ten to twenty years and that plastics packaging is a major contributor to the use of these reserves. There are two parts to any answer to this argument. The first is that oil reserves quoted are proven oil reserves and do not take account of sources, such as the Athabasca Tar Sands, which would double the figures if they were included. The reason they are not included is that they are, at present, uneconomic to exploit, but if other reserves were used up, the picture would change radically in their favour. In addition, fresh oil reserves are being found yearly. Long before the time that oil reserves really are dangerously low other fuel sources such as atomic power will be contributing a major share to supplying the total energy demand. The other part of the answer is that only 1 per cent of the world's yearly consumption of oil is used as a feedstock for plastics manufacture. Even more significant, only a quarter of the total plastics production ends up as packaging. In other words, if oil reserves were to sink to ten years' supply (based on their use as fuel) before other energy sources took over, they would last for 4,000 years as a supply for plastics packaging.

Plastics can also, of course, be made from other materials, notably coal and cellulose and work has also been carried out on making plastics from sugar.

Tinplate The tin is now only a minor component of tinplate and is still being reduced in quantity. The steel component is mainly dependent on supplies of iron ore and these are still plentiful. In addition, scrap iron is re-cycled and the percentage so used could be increased when it became economically essential.

239

Conclusions

1　Packaging does not present an insuperable problem to local authorities provided that they are kept well-informed of trends in packaging to enable them to plan future disposal techniques.
2　Disposability must be included in packaging development plans. Current packs should be re-examined for fitness of purpose (has over-packaging crept in?). There may be possibilities for cost savings.
3　More attention should be paid to public relations in this field.

13

Education and Training in Packaging

The importance of the packaging function is becoming widely recognised in industry and this is being followed by a growing awareness of the importance of packaging education and training.

Need for education

It is no longer possible for the seller of packaging to be just an order taker. Also, is it not feasible for the package buyer to order, say, glass bottles for a liquid product from a particular supplier, without taking into account not only glass bottles from a competitive supplier, but alternative materials, such as plastics bottles, lined cartons, plastics pouches, etc. One example of this is the field of liquid milk packaging. Once, the dairyman need only consider buying glass bottles and was usually content to re-order from his normal supplier with no change in the type of bottle. A minimum knowledge of 'packaging' was necessary and so the need for education was slight. Then came developments in the light-weighting of glass bottles together with some new shapes designed to cut down abrasion of bottles during handling by reducing the possible area of contact between one bottle and the next.

These new bottles had to be assessed for behaviour during filling and during distribution and an awareness of the subject of 'packaging' was born. Nowadays, there are many different possible packages for liquid milk, including cartons, (of various types), plastics film sachets, and plastics bottles. The latter are again sub-divided into those which are made, stored and then filled, and those which are made and filled in-plant on the same machine. The advantages and disadvantages of these widely different packages depend on a number of factors which can only really be evaluated by the particular dairy. These include the number of trips normally averaged by the existing glass bottle, the cost of the new package, any savings which might be made on the round due to the lighter weight and lesser volume of the new packages, cost of new filling plant, filling plant speeds, and the consumer's acceptance or otherwise of the new package. The dairyman is thus forced to learn a great deal about packaging if he is to make the right choice for his particular situation. It is not possible for him to blindly copy what a neighbouring dairy has done; he must carry out his own investigations.

Similarly, the supplier of packaging. The glass bottle supplier now has to be *au fait* with a number of competing systems if he is to convince his customer that glass is still the best pack for his needs.

Most people will agree with these arguments and yet some companies still build-up their packaging function in a haphazard manner. In some companies the packaging manager will have been in the buying department (not even package buying necessarily) until he was transferred to the packaging department, in another company it may be a research and development man who gets the job. There are still packaging suppliers who recruit technical representatives from men with possibly a chemical or engineering background, but with no knowledge of the packaging industry. It is becoming more and more difficult to obtain suitable personnel by these methods and packaging education and training can no longer be ignored.

The importance of education and training to all sides of the packaging industry can be summarised as follows:

1 Users
(a)   Rapid growth of new packaging materials makes it essential to keep well-informed of their properties in order to obtain better package performance at the same cost or the same package performance at lower cost.

(b)   New developments in packaging equipment must be assessed. In addition to the possibility of faster filling speeds or lower packaging costs, through new filling equipment, there are often developments in in-plant manufacture of packages to be considered.

(c)   A knowledge of packaging economics is essential to fully assess advances in both materials and equipment.

2   Material suppliers

(a)   Packaging material manufacturers must have a knowledge of packaging equipment requirements otherwise expensive developments on materials can be lost because the materials are not suitable for existing packaging equipment.

(b)   Packaging material manufacturers must also have a wide knowledge of their end-use markets and trends. Unsatisfactory packages for a particular product will not lead to repeat sales.

3   Equipment suppliers

Equipment suppliers must keep abreast of advances in packaging materials in order to take advantage of new markets (or hold on to old ones).

Packaging education

The need for packaging education is being met in varying degrees by a variety of methods and by organisations both in the UK and abroad. An outline of the major schemes available is given below.

United Kingdom

The qualifying body for packaging in the UK is the Institute of Packaging. The grades of membership are as follows:

1   Fellows
2   Members
3   Associates
4   Students

An examination is held annually for Membership while Fellowships are awarded to those who have been Members for five years and who, in the opinion of the Fellowship Credentials Panel have made a positive contribution to packaging technology. The Fellowship Credentials Panel normally conducts an oral examination for candidates.

Associateship is open to anyone who is directly concerned with the packaging industry but is not eligible for election as a Member or to those who are interested but not directly concerned with packaging.

Student membership is a prerequisite for sitting the Institute's Membership examination. The examination itself consists of three papers of three hours duration each, plus a compulsory essay on a subject set three months before the written examination (the date on which the essay has to be handed in). The essay, which must be not less than 2,000 words, is not designed to test what the candidate already knows, but to test his or her ability to search out information and draw logical conclusions concerning the selected subject.

In the written question papers there is a wide choice given in each paper to allow for the fact that packaging covers such a very wide field. It is also possible to be referred in certain papers (including the essay) so that only those papers need be taken at any second attempt.

The syllabus for the examination is too long to be given here, in full, but the various headings will give some idea of the scope covered. Thus the syllabus covers the following:

> Introduction to packaging
> The necessity for packaging
> Principles of protection
> Packaging materials
> Types of packaging
> Accessories for packaging
> Packaging and production processes
> Packed goods
> Materials handling, movement and storage
> Package testing and development
> Specifications and quality measurement for control
> Rationalisation and standardisation
> Legal requirements
> Marking, identification and labelling
> Package design
> Economics of packaging
> Applied packaging (including packaging of foodstuffs, pharmaceuticals, chemicals, engineering equipment, etc).

Examination questions in the past have included questions on the fundamentals of packaging, such as journey hazards, adhesives, cushioning, permeability of packaging material and packaging criteria, together with applied packaging, printing and labelling, packaging machinery, and distribution. A good example of the type of essay set, is given below.

'For many years packaging standards have been based on specifications for materials. There is now a move towards producing specifications based on performance requirements of materials, containers and finished packs. Compare the merits of the two systems and discuss the technical problems involved in producing performance specifications for *one* of the following examples:

(a)   Printed packs for 8 oz of a processed foodstuff to be filled on a high speed line.

(b)   Containers for a cosmetic cream.

(c)   Outer containers for A1 Tall cans.

More than one type of container can be used in each example. In your chosen example, list the properties which need to be specified for each of the several possible containers.'

Courses are run at various colleges, in London and elsewhere, which are designed to help the student to pass the Institute's examination. Courses are usually mounted over a period of two years, one evening a week, during two terms a year of about twelve weeks each. The lecturers at these courses are normally experts from industry, lecturing on their specialist subjects. Students are expected to supplement these lectures by extensive reading of the technical press, and the two text books, published under the authority of the Institute of Packaging Council, one on Packaging Fundamentals and the other on Packaging Materials and Containers.

In addition, the Institute runs a correspondence course in conjunction with the National Extension College. This has the advantage that the student is given questions to answer at the end of each of the twenty–six lessons and these are marked and commented on by individual tutors.

In addition to these activities, which are aimed at preparing students for the examination, the Institute also runs one and two week residential courses at universities such as Reading, Nottingham and Sussex, on general or specialised packaging topics. The latter have included, in the past, such topics as Food Packaging and the Packaging of Pharmaceuticals and Cosmetics. Other activities include two and three day conferences and one day symposia on topics such as 'Developments in shrinkwrapping', 'What's new in packaging foams?' and 'Packaging for profit'. Finally, of course, there are the branch evening meetings mentioned in Chapter One.

There is no one Training Board which covers the packaging industry but Boards having an interest in packaging include those covering Paper and Paper Products, Food, Drink and Tobacco, and Chemicals and Allied Products. In addition, PIRA (the Research Association for the Paper and Board, Printing and Packaging Industries) runs training courses for machine operators, technicians and management on a wide range of packaging topics. Some past courses have included 'Choosing and checking fibreboard packaging', 'Physical testing of paper and board' and 'Letterpress and lithographic trouble-shooting — a guide to techniques and methodology'.

The larger companies have their own facilities for training although even they may obtain specialist advice from outside.

North America

As one would expect from the land which was the cradle of modern packaging, there are a large number of courses covering a wide variety of topics. In the USA, a number of colleges and universities run packaging courses ranging from specialised evening ones such as a fifteen week course on Cosmetics Packaging at Columbia University, New York and a fifteen week course on Industrial Packaging, at Indiana University, Indianapolis, to the more general type of course such as Packaging, Packaging Materials, offered by Rochester Institute of Technology, Rochester, New York.

The American Management Association holds a series of seminars and courses in New York and Chicago throughout the year, while New York University also offers an evening course covering various aspects of Packaging Management and Technology. Rather more unusual types of courses are offered by the Joint Military Packaging Training Centre, Aberdeen Proving Grounds, Maryland. These are free and are for employees of companies holding Government contracts, supplying packaging materials, or offering packaging services to the military services. Titles of courses include Packaging Design, Preservation and Intermediate Protection, Packaging and Containerisation, Basic Packing, Preservation and Packaging, Equipment Preservation for Shipment or Storage, Preparation of Freight for Air Shipment, and Inspection of Packaged and Packed Household Goods for Storage and Shipment.

There are also a number of full-time undergraduate and graduate courses, the best known being those available at Michigan State University, East Lansing. This University has a School of Packaging running undergraduate and graduate programmes leading to BS and MS degrees. Courses include subjects such as Packaging Machinery, Packaging Design, Principles of Packaging, Packaging Materials, Packaging Systems, Packaging Operations, Packaging Analysis, Dynamics of Packaging and Packaging Developments.

In Canada, the Packaging Association of Canada runs a Packaging Course which takes up one full day a month. In addition, it runs training courses for packaging mechanics. One interesting project is that aimed at teaching developing nations the basics of packaging, the course being based on tape recordings allied to visual aids.

Europe

One big difference between the UK and the rest of Europe is that the British Institute of Packaging is the only one based on individual membership and having an entrance examination. Other European Institutes are either Government run or based on Corporate Membership. In Holland, for example, the Nederland Verpackingscentrum (Netherlands Packaging Centre) is an association of companies, including those making packaging materials, packages, packaging equipment, companies using packages and packaging materials, transport companies, research institutes and advertising agencies, designers and packaging specialists. It organises regional meetings, industrial visits and national events such as Congresses, and an Export Packaging Day.

In conjunction with the TNO Institute for Packaging Research and the Education and Training Department of the Netherlands Institute for Efficiency, the Netherlands Packaging Centre also organises packaging courses. These include a General Basic Packaging Course; Paper, Cardboard, and Plastics Films for Retail Packages; Cardboard Packaging for Transport; Plastics for Packaging; and Creativity and Packaging.

In France, the French Institute of Packaging runs both general and specialised seminars and courses, ranging from one and two-day events to those of two weeks duration. The French Institute has also purchased the Canadian tape recording/visual aid course and is adapting it to the needs of French-speaking developing countries.

The World Packaging Organisation (formed from the various Continental Federations of Packing Institutes) has also recognised the need for packaging training. A common blue-print, based on the British Institute of Packaging syllabus has been prepared and is now in use internationally.

Packaging education, then, is still expanding in an effort to keep pace with increasing demand. People's needs are individual and this book has been written with the needs of management firmly in mind. It is hoped, therefore, that it will prove useful to the manager involved in the growing complexity of the Packaging Function.

Index

249